'O' Level Arithmetic in One Year

ISBN 0 7169 6937 8

© 1974, George McCall.

All rights reserved. No part of this publication may be reproduced; stored in a retrieval system; or transmitted in any form or by any means — electronic, mechanical, photocopying, recording or otherwise — without the prior permission of the publisher, Robert Gibson & Sons, Ltd., 17 Fitzroy Place, Glasgow G3 7SF.

'O' Level Arithmetic in One Year

SECOND EDITION

GEORGE McCALL

ROBERT GIBSON · PUBLISHER
17, Fitzroy Place, Glasgow, G3 7SF.

This text book is mainly intended for pupils in 'arithmetic only' sections or mature students attending Further Education Colleges with the intention of taking an Arithmetic 'O' Grade in one year.

If one exercise is completed each day during a school session then the pupil or student will have covered most of the topics needed in the S.C.E. syllabus, with time to spare for revision purposes. Explanations and worked examples are included only where it was felt necessary and the examples are graded in degree of difficulty.

Pupils who are on a bridge course can spread the exercises over a two year period and there are sufficient simpler examples on each topic to provide a suitable course for non-certificate pupils.

CONTENTS

		Page
Chapter 1	The four rules;	7
Chapter 2	Factors fractions and ratios;	13
Chapter 3	Percentages;	19
Chapter 4	Standard form; significant figures and decimal places;	22
Chapter 5	Square roots;	24
Chapter 6	Sequences; Binary numbers; number bases;	29
Chapter 7	The metric system	35
Chapter 8	Direct and inverse proportion. Averages and speed averages;	39
Chapter 9	Profit and Loss per cent. Simple interest. Compound interest and increases. Discounts and commissions;	44
Chapter 10	Insurance. Assurance. Bankruptcy. Foreign exchange	56
Chapter 11	Rates. Income tax. Gas and electricity bills;	62
Chapter 12	Logarithms;	71
Chapter 13	Areas;	80
Chapter 14	Circumference and area of circle.	89
Chapter 15	Volumes of various solids;	94
Chapter 16	Stocks and Shares;	101
Chapter 17	Estimation of Errors;	105
Chapter 18	Probability. Tree diagrams and Permutations;	109
Chapter 19	Statistics;	115
Answers		151

Printed by Robert McLehose & Co., Ltd., Glasgow.

CHAPTER 1 The four rules on numbers, money, weight, length and capacity.

EXERCISES 1 – 14

Exercise 1
Add the following:

1. 6951
   2714
   645
   9273
   96
   ———

2. 36475
   2895
   3384
   9376
   893
   ———

3. 72003
   53886
   708
   894
   6593
   ———

4. 653
   2402
   5632
   78669
   7429
   ———

5. 18693
   4172
   7402
   10964
   4327
   ———

6. 9321
   3478
   27195
   7836
   38674
   ———

7. 27639
   41263
   20739
   7485
   87
   ———

8. 57964
   75072
   10986
   369
   81652
   ———

9. 8643
   793
   5062
   39507
   6728
   ———

10. 8421
    7124
    37456
    159
    4317
    ———

Exercise 2
Subtract:

1. 47938
   28794
   ―――

2. 8724
   6597
   ―――

3. 4783
   3202
   ―――

4. 97352
   43874
   ―――

5. 82613
   78704
   ―――

6. 8522
   6395
   ―――

7. 95330
   41852
   ―――

8. 9274
   9186
   ―――

9. 40877
   38749
   ―――

10. 96405
    67127
    ―――

Exercise 3
Simplify:

1. 2158 − 538 + 496 + 728 − 1587
2. 3763 + 4384 − 6429
3. 763 − 689 − 345 + 472
4. 2876 − 576 + 298
5. 634 + 5627 − 4893
6. 4736 + 87 − 1272 − 375
7. 2017 − 1207 − 225 + 96
8. 2894 + 2896 − 755 − 2704
9. 4718 − 642 − 763 + 289
10. 5864 − 386 + 215 − 4281

Exercise 4
Multiply the following numbers together:
1. 563 x 28
2. 3762 x 49
3. 8047 x 52
4. 365 x 24
5. 968 x 165
6. 8142 x 326
7. 954 x 372
8. 4683 x 89
9. 2685 x 96
10. 1197 x 427

Exercise 5
Find the quotient and remainder in the following examples:
1. 3795 ÷ 48
2. 7693 ÷ 54
3. 8697 ÷ 37
4. 9602 ÷ 69
5. 17368 ÷ 204
6. 37692 ÷ 315
7. 38784 ÷ 425
8. 57894 ÷ 236
9. 28459 ÷ 392
10. 76359 ÷ 471

Exercise 6
Add together:
1. 7.65 + 8.09 + 24.68 + 17.91 + 5.73
2. 16.72 + 34.173 + 5.106 + 68.32 + 7.154
3. 2.86 + 0.92 + 30.616 + 4.875 + 14.627
4. 0.782 + 5.816 + 4.36 + 3.001 + 12.483
5. 0.827 + 2.94 + 0.203 + 3.102 + 0.372
6. 37.04 + 0.892 + 7.364 + 9.197 + 6.14
7. 3.65 + 91.6 + 0.808 + 7.247 + 0.134
8. 86.155 + 4.731 + 21.53 + 3.907 + 7.13
9. 3.706 + 0.587 + 0.084 + 2.07 + 3.459
10. 5.68 + 2.927 + 0.692 + 3.958 + 6.530

Exercise 7
Subtract:
1. 10.89 − 3.61
2. 3.481 − 2.359
3. 0.8649 − 0.1195
4. 61.461 − 52.903

contd. overleaf

5. 7.842 − 2.916
6. 0.4096 − 0.1576
7. 3.924 − 3.098
8. 8.496 − 5.937
9. 16.089 − 13.263
10. 4.362 − 3.806

Exercise 8
Simplify:
1. 3.96 + 2.87 + 7.09 − 3.54
2. 72.07 + 63.8 + 2.94 − 48.43
3. 8.54 − 7.18 + 21.73 + 15.82 − 27.05
4. 86.39 + 97.06 + 79.28 − 65.07 − 82.06
5. 7.518 + 1.106 + 1.804 − 9.003
6. 7.422 + 8.623 + 15.075 − 24.918
7. 3.706 + 10.084 − 5.369 + 1.055
8. 2.831 + 1.009 − 2.968 + 0.573
9. 2.184 + 6.255 + 14.062 − 20.731
10. 95.043 + 17.834 + 12.046 − 34.528

Exercise 9
Multiply together the following numbers:
1. 5.36 x 1.24
2. 36.6 x 3.7
3. 45.7 x 4.26
4. 6.92 x 2.08
5. 3.62 x 3.65
6. 84.2 x 5.7
7. 18.5 x 0.49
8. 0.952 x 1.32
9. 63.4 x 2.53
10. 7.62 x 0.57

Exercise 10
Division of Decimals. Divide the following numbers:
1. 9.6 ÷ 0.016
2. 7.314 ÷ 2.3
3. 46.5624 ÷ 232
4. 0.08316 ÷ 9
5. 0.738 ÷ 0.006
6. 0.37236 ÷ 1.2
7. 0.0315 ÷ 0.45
8. 23.6081 ÷ 470
9. 364.77 ÷ 0.21
10. 17.355 ÷ 0.65

Exercise 11
Addition, Subtraction, Multiplication and Division of Compound Quantities:

Add:
1. £7.73 + £9.63 + £9.67 + £5.44
2. £18.64 + £6.67 + £5.93 + £9.26 + £9.46
3. £61.88 + £78.16 + £7.74 + £67.02 + £8.65
4. £9.28 + £39.23½ + £6.75½ + £13.28 + £24.68½
5. £7.43 + £8.18½ + £42.83 + £26.91½ + £83.10½
6. 8.56 kg + 9.37 kg + 6.91 kg + 5.88 kg
7. 96.17 kg + 17.98 kg + 81.46 kg + 46.25 kg
8. 6.515 kg + 3.563 kg + 16.85 kg + 5.881 kg
9. 4.265 kg + 5.761 kg + 11.819 kg + 10.716 kg
10. 0.658 kg + 0.861 kg + 0.116 kg + 0.459 kg
11. 18.63 l + 1.75 l + 24.50 l + 9.12 l
12. 7.321 l + 0.864 l + 1.607 l + 0.818 l
13. 2.370 l + 4.371 l + 7.4 l + 5.392 l
14. 52.45 m + 378.40 m + 20.86 m + 87.32 m
15. 14.70 m + 52.96 m + 3.142 m + 2.25 m

Exercise 12
Subtract:
1. £817.59½ − £789.76
2. £1051.75 − £659.91½
3. £2000 − £958.82
4. £952.79 − £786.94½
5. £643.74½ − £568.27½
6. 85.632 kg − 47.091 kg
7. 2.064 kg − 1.764 kg
8. 84.57 kg − 37.419 kg
9. 11.717 kg − 3.702 kg
10. 12.506 kg − 7.373 kg
11. 37.38 l − 26.92 l
12. 7.408 l − 6.153 l
13. 197.32 m − 83.45 m
14. 4.762 m − 3.904 m
15. 67.087 m − 32.045 m

Exercise 13
Multiply:
1. £82.34 x 9
2. £37.46 x 12
3. £1.17 x 18
4. £3.29 x 24
5. £8.66½ x 29
6. 7.34 kg x 14
7. 12.69 kg x 23
8. 14.98 kg x 45
9. 13.62 kg x 56
10. 26.32 kg x 125
11. 3.604 l x 21
12. 9 64 l x 15
13. 3.78 m x 64
14. 8.98 m x 26
15. 48.23 m x 87

Exercise 14
Divide:
1. £26.34 ÷ 6
2. £85.82 ÷ 7
3. £3097.92 ÷ 12
4. £192.92 ÷ 53
5. £146.28 ÷ 23
6. £4285.02 ÷ 17
7. 2.673 kg ÷ 3
8. 36.54 kg ÷ 29
9. 9.168 kg ÷ 8
10. 143.28 kg ÷ 6
11. 38.988 l ÷ 27
12. 25.488 l ÷ 18
13. 192.128 m ÷ 16
14. 157.5 m ÷ 45
15. 47.2162 m ÷ 47

CHAPTER 2 Factors, Fractions, Operations on Fractions, Fractions of Concrete Quantities and Ratios.

EXERCISES 15 — 28

Exercise 15

All positive integral numbers are either **prime** or **composite**.

A **prime** number is of the type 2, 3, 5, 11, 19. It can only be divided by itself.

A **composite** number is the product of two or more other numbers.

$$60 = 4 \times 15 = 5 \times 12 = 6 \times 10 \text{ etc.}$$

The numbers 4, 15, 5, 12 . . . are **factors** of 60.

A composite number is often expressed in **prime factor** form.

$$60 = 2 \times 2 \times 3 \times 5$$

Break up into Prime Factors:

1. 140
2. 315
3. 308
4. 364
5. 504
6. 145
7. 378
8. 147
9. 1287
10. 792

Exercise 16

Lowest Common Multiple (L.C.M.) This is the smallest composite number into which two or more other numbers will divide.

It may be obtained by inspection or by breaking the given numbers into prime factors.

Example 1 Find L.C.M. of 5, 9 and 14.

$$5 = 5$$
$$9 = 3 \times 3$$
$$14 = 2 \times 7$$
$$\text{L.C.M.} = 2 \times 3 \times 3 \times 5 \times 7 = 630$$

Example 2 Find L.C.M. of 6 and 32

$$6 = 2 \times 3$$
$$32 = 2 \times 2 \times 2 \times 2 \times 2 = 2^5$$
$$\text{L.C.M.} = 2^5 \times 3 = 96$$

Find L.C.M. of
1. 8, 9 and 12.
2. 12 and 16.
3. 2, 6 and 16.
4. 18 and 45.
5. 8, 12 and 28.
6. 8, 9 and 15.
7. 5, 7 and 9.
8. 6, 8 and 14.
9. 7, 21 and 28.
10. 5, 11 and 35.

Exercise 17

Reduce the following fractions to their simplest terms.

1. $\frac{6}{9}$
2. $\frac{24}{60}$
3. $\frac{15}{25}$
4. $\frac{28}{84}$
5. $\frac{18}{54}$

6. $\frac{128}{240}$
7. $\frac{72}{108}$
8. $\frac{50}{175}$
9. $\frac{26}{143}$
10. $\frac{88}{132}$

Exercise 18

Express the following improper fractions as mixed numbers.
Example: $\frac{26}{9} = 2 + \frac{8}{9} = 2\frac{8}{9}$

1. $\frac{8}{3}$
2. $\frac{15}{8}$
3. $\frac{26}{7}$
4. $\frac{90}{13}$
5. $\frac{27}{6}$

6. $\frac{28}{11}$
7. $\frac{36}{7}$
8. $\frac{14}{5}$
9. $\frac{13}{4}$
10. $\frac{71}{15}$

Exercise 19

Express the following mixed numbers as improper fractions.
Example: $6\frac{3}{8} = \frac{48}{8} + \frac{3}{8} = \frac{51}{8}$

1. $3\frac{3}{4}$
2. $4\frac{1}{5}$
3. $6\frac{2}{9}$
4. $5\frac{3}{7}$
5. $7\frac{2}{3}$

6. $9\frac{1}{7}$
7. $10\frac{1}{2}$
8. $11\frac{5}{8}$
9. $3\frac{5}{9}$
10. $16\frac{1}{3}$

Exercise 20
Addition of Fractions:

Example: $3\frac{1}{5} + 5\frac{9}{10} = \frac{16}{5} + \frac{59}{10} = \frac{32}{10} + \frac{59}{10} = \frac{91}{10} = 9\frac{1}{10}$

Simplify the following:

1. $\frac{1}{2} + \frac{5}{6}$
2. $\frac{2}{3} + \frac{7}{8}$
3. $\frac{3}{4} + \frac{2}{5}$
4. $\frac{2}{3} + \frac{1}{4} + \frac{3}{5}$
5. $\frac{2}{5} + \frac{9}{10} + \frac{7}{15}$
6. $2\frac{1}{6} + 1\frac{3}{4}$
7. $4\frac{5}{9} + 2\frac{11}{15}$
8. $4\frac{5}{8} + 5\frac{3}{4} + 6\frac{7}{12}$
9. $3\frac{5}{6} + 2\frac{3}{4} + 6\frac{1}{8}$
10. $3\frac{2}{3} + 2\frac{1}{6} + 5\frac{8}{9}$

Exercise 21
Subtraction of Fractions:

Example: $6\frac{7}{9} - 3\frac{5}{6} = \frac{61}{9} - \frac{23}{6} = \frac{122}{18} - \frac{69}{18} = \frac{53}{18} = 2\frac{17}{18}$

1. $\frac{13}{15} - \frac{7}{10}$
2. $\frac{7}{8} - \frac{5}{12}$
3. $8\frac{5}{6} - 4\frac{3}{8}$
4. $4\frac{5}{7} - 2\frac{3}{14}$
5. $5\frac{1}{2} - 4\frac{7}{8}$
6. $6\frac{8}{21} - 3\frac{11}{14}$
7. $8\frac{6}{25} - 5\frac{4}{5}$
8. $3\frac{7}{8} - 1\frac{5}{6}$
9. $7\frac{8}{9} - 4\frac{7}{12}$
10. $8\frac{4}{7} - 6\frac{2}{3}$

Exercise 22
Simplify the following:

Example: $6\frac{1}{4} - 3\frac{7}{8} + 2\frac{1}{2} = \frac{25}{4} - \frac{31}{8} + \frac{5}{2} = \frac{50}{8} - \frac{31}{8} + \frac{20}{8} = \frac{39}{8} = 4\frac{7}{8}$

1. $5\frac{5}{6} + 4\frac{2}{3} - 6\frac{11}{12}$
2. $5\frac{2}{5} + 3\frac{1}{3} - 6\frac{8}{15}$
3. $6\frac{7}{12} - 7\frac{3}{8} + 3\frac{11}{16}$
4. $10\frac{7}{16} + 5\frac{7}{24} - 13\frac{1}{4}$
5. $3\frac{1}{2} + 4\frac{2}{3} + 2\frac{1}{4} - 6\frac{5}{6}$
6. $3\frac{6}{7} - 7\frac{3}{28} + 8\frac{11}{14}$
7. $6\frac{3}{4} - 5\frac{1}{3} + 2\frac{7}{8}$
8. $4\frac{7}{12} - 3\frac{5}{6} + 6\frac{4}{9}$
9. $5\frac{2}{3} + 6\frac{1}{2} + 3\frac{5}{8} - 10\frac{1}{4}$
10. $7\frac{3}{4} - 4\frac{10}{11} + 1\frac{9}{44}$

Exercise 23
Multiplication of Fractions:

Example: $3\frac{3}{5} \times 2\frac{1}{2} = \frac{\cancel{18}^{9}}{\cancel{5}_{1}} \times \frac{\cancel{5}^{1}}{\cancel{2}_{1}} = 9$

Simplify the following:

1. $4\frac{2}{7} \times 2\frac{1}{3}$
2. $6\frac{2}{3} \times 1\frac{1}{5}$
3. $\frac{8}{15} \times 3\frac{3}{4}$
4. $8\frac{1}{3} \times 2\frac{14}{15} \times 1\frac{7}{11}$
5. $6\frac{12}{13} \times \frac{1}{8} \times 1\frac{5}{21}$
6. $4\frac{1}{6} \times 6\frac{2}{5}$
7. $\frac{4}{11} \times 9\frac{3}{7} \times 1\frac{1}{2}$
8. $\frac{6}{7} \times 1\frac{2}{5} \times \frac{2}{3}$
9. $3\frac{4}{7} \times 4\frac{2}{3} \times \frac{18}{25}$
10. $1\frac{4}{9} \times 1\frac{13}{14} \times \frac{7}{13}$

Exercise 24
Division of Fractions:

Example: $3\frac{1}{9} \div \frac{7}{12} = \frac{3\frac{1}{9}}{\frac{7}{12}} = \frac{\frac{28}{9} \times \frac{12}{7}}{\frac{7}{12} \times \frac{12}{7}} = \frac{\frac{16}{3}}{1} = 5\frac{1}{3}$

This may be shortened to:

$$3\frac{1}{9} \div \frac{7}{12} = \frac{28}{9} \times \frac{12}{7} = \frac{16}{3} = 5\frac{1}{3}$$

Simplify the following:

1. $1\frac{7}{8} \div \frac{9}{12}$
2. $10\frac{2}{3} \div 1\frac{1}{9}$
3. $4\frac{2}{7} \div 9\frac{9}{14}$
4. $\frac{6}{7} \div \frac{3}{14}$
5. $14\frac{2}{5} \div 2\frac{7}{10}$
6. $12\frac{2}{5} \div 3\frac{1}{10}$
7. $8\frac{3}{11} \div 8\frac{2}{3}$
8. $6\frac{4}{5} \div 7\frac{1}{2}$
9. $3\frac{2}{7} \div 2\frac{7}{8}$
10. $9\frac{3}{4} \div 8\frac{2}{3}$

Exercise 25
Mixed Examples:

Simplify **"Brackets"** and **"of"** first followed by multiplication and division and finally add or subtract the remaining terms.

1. $\frac{3}{4} + \frac{1}{2} \times \frac{4}{7}$
2. $(\frac{3}{8} + \frac{1}{2}) \times \frac{4}{9}$
3. $\frac{3}{7} \times \frac{4}{5} \div 4\frac{1}{5}$
4. $2\frac{4}{5} \times 1\frac{7}{9} + 3\frac{2}{15}$
5. $(4\frac{1}{2} + 3\frac{1}{3}) \div (3\frac{2}{3} - 1\frac{1}{2})$
6. $3\frac{3}{4} - \frac{2}{3}$ of $1\frac{5}{7}$
7. $3\frac{2}{3} - 2\frac{1}{2} \div 1\frac{12}{13} + 1\frac{1}{2}$
8. $\frac{1}{5} \div 1\frac{2}{3} + 2\frac{7}{10} \div 7\frac{1}{5}$
9. $(2\frac{3}{4} - 1\frac{2}{3}) \div (\frac{1}{2} + \frac{1}{3})$
10. $4\frac{5}{6} - (1\frac{2}{3}$ of $4\frac{2}{5}) + 6\frac{1}{4}$
11. $(3\frac{1}{2} - 2\frac{5}{8}) \div (4\frac{1}{4} - 3\frac{2}{3})$
12. $5\frac{1}{3} - 5\frac{1}{2} \div 4\frac{2}{5}$
13. $(3\frac{5}{6} + 1\frac{1}{2})$ of $(3\frac{1}{8} - 2\frac{3}{4})$
14. $3\frac{2}{3} - (4\frac{3}{8} \div 3\frac{1}{2})$
15. $1\frac{2}{3}$ of $4\frac{1}{2} - 5\frac{1}{8}$

Exercise 26   Fractions of Concrete Quantities

1. Find the value of 2¼ times £273.36.
2. Find the value of $\frac{3}{7}$ of £9.52.
3. 5½ litres of paraffin at 3p per litre.
4. 16¼ metres of wire at 20p per metre.
5. 2¾ kg. of flour at 8p per kg.
6. 1½ kg. of margarine at 5½p per quarter kg.
7. 3¾ kg. of sugar at 10p per kg.
8. 6½ metres of cloth at 69p per metre.
9. 27¾ metres of curtain material at £1.24 per metre.
10. Find $\frac{3}{8}$ of 84 litres.
11. Find the value of 1¾ of 46 quintals.
12. Find the value of $2\frac{1}{3}$ of £2.67.
13. $\frac{3}{20}$ of 175 kg.
14. 14½ litres of milk at 12p per litre.
15. 9½ litres of paint at £1.26 per litre.

Exercise 27

Ratio: The ratio of two quantities is the fraction that the one is of the other. The quantities in a ratio must be of the same name and the same unit.

Example: Find the ratio of £1.50 to 36p.

$$\text{Ratio} = \frac{\cancel{150}^{75}}{\cancel{36}_{18}} = \frac{\cancel{75}^{25}}{\cancel{18}_{6}} = \frac{25}{6}$$

Find the ratio of:
1. 27p to 84p
2. 6.5 to 7.25
3. 57 metres to 42 metres
4. £3.75 to £8.50
5. $4\frac{1}{8}$ to $6\frac{7}{8}$
6. 6 days to 1 day 8 hours
7. 645 gm to 3 kg
8. £1.89 to £8.82
9. 6200 cm² to 1 m²
10. 850 c.c.s. to 1.5 litres

Exercise 28

1. What sum of money bears to 75p the ratio 3 : 5?
2. What area bears to 176 cm² the ratio 8 : 11?
3. What weight bears to 56 kg the ratio 5 : 8?
4. What volume bears to 2.7 litres the ratio 7 : 9?
5. What sum bears to £8.46 the ratio the ratio 7 : 3?
6. The volume 39 litres bears to a second volume tha ratio 13 : 16; find the second volume.
7. The weight 14.3 gm bears to a second weight the ratio 11 : 2; find the second weight.
8. The area 72 m² bears to a second area the ratio 16 : 9; find the second area.
9. The length 7.5 metres bears to a second length the ratio of 5 : 8 find the second length.
10. £30.80 bears to a second sum of money the ratio of 7 : 20; find the second sum of money.

CHAPTER 3. Percentages. Changing to fractions. Fractions to percentages. Percentages of concrete quantities. One quantity as the percentage of another.

EXERCISES 29 – 33

Exercise 29
Percentages. A percentage is a fraction in which the denominator is 100.

Example: $\quad 65\% = \dfrac{\cancel{65}^{13}}{\cancel{100}_{20}} = \dfrac{13}{20}$

Express the following percentages as fractions in their lowest terms.

1.	70%	2.	22%	3.	36%	4.	45%
5.	75%	6.	84%	7.	115%	8.	40%
9.	28%	10.	95%	11.	62½%	12.	12½%
13.	2½%	14.	37½%	15.	$33\frac{1}{3}$%	16.	$66\frac{2}{3}$%
17.	87½%	18.	6¼%	19.	$9\frac{3}{8}$%	20.	11¼%

Exercise 30
To express a fraction as a percentage multiply by 100.

Example: $\quad$ Express $\dfrac{9}{16}$ as a percentage

$$\dfrac{9}{16} = \dfrac{9}{\cancel{16}_4} \times \dfrac{\cancel{100}^{25}}{1} \text{ per cent} = \dfrac{225}{4} = 56\tfrac{1}{4}\%$$

Example: $\quad$ Express 0.342 as a percentage

$\quad\quad\quad\quad$ 0.342 = 0.342 × 100 per cent

$\quad\quad\quad\quad\quad\quad\,\,\,\,$ = 34.2%

Express the following fractions and decimals as percentages.

1. $\dfrac{3}{4}$ $\quad\quad$ 2. $\dfrac{5}{8}$ $\quad\quad$ 3. $\dfrac{5}{16}$ $\quad\quad$ 4. $2\dfrac{1}{5}$

5. $\frac{2}{3}$ 6. $1\frac{3}{4}$ 7. $\frac{7}{25}$ 8. $\frac{3}{5}$
9. $1\frac{1}{3}$ 10. $\frac{17}{25}$ 11. 0.5   12. 0.25
13. 0.625   14. 5.25   15. 0.375   16. 1.25
17. 0.0075   18. 2.50   19. 0.125   20. 0.0875

Exercise 31
Percentages of Concrete Quantities:
Example: Find the value of 2½% of £120.

$$= \frac{5}{200} \times £120 = \frac{1}{40} \times £120 = £3$$

Find the value of:
1. 12% of £17.50
2. 10% of £11.25
3. 62½% of £41.36
4. 75% of £77.20
5. 12½% of £97
6. $12\frac{4}{5}$% of 150 gm
7. $26\frac{2}{3}$% of 1.32 litres
8. 26¼% of 3.5 kg
9. 22½% of 7.5 km
10. $57\frac{1}{7}$% of 1 hr 17 min
11. 215% of 1120 metres
12. 15% of 2.05 kg
13. 4% of 27.5 gm
14. $3\frac{1}{11}$% of 14.5 litres
15. 18¾% of 8.192 km

Exercise 32
Example: What percentage is 550 metres of 11 km?

$$\frac{5\cancel{50}}{11{,}\cancel{000}} \times \frac{1\cancel{00}}{1} \text{ per cent} = 5\%$$

What percentage is:
1. 75p of £2.25?
2. 3.5 metres of 18 metres?
3. 660 of 1760?
4. 90p of £1.50?
5. 36cm of 3 metres?
6. $6\frac{1}{8}$ days of 1 week?
7. £2.52 of £10.50?
8. 800 cm^2 of 1 m^2?
9. 950 gm of 2.5 kg?
10. 280 m of 7 km?

Exercise 33

1. If 25% of my money is 43p, how much have I?
2. 6¼% of my salary is £70. How much do I earn?
3. 650 pupils sat a common examination and 8% failed. How many passed?
4. The difference between 15% of a certain volume and 25% is 42 litres. What is the total volume?
5. If 70% of my money is £4.20, how much do I have?
6. Find the weight of which 8% is 3.36 kg.
7. 56% of a certain length is 84 cm. What is the total length?
8. What is my income when 13% of it is £416?
9. 8% of the area of a farm is 56 hectares. What is the total area of the farm?
10. 35% of the cost of a suite is required as a deposit. If the deposit is £52.50, what is the total value of the suite?

CHAPTER 4. Numbers in standard form or scientific notation. Significant figures and prescribed numbers of decimal places.

EXERCISES 34 − 36

Exercise 34
Numbers in standard form or scientific notation:
A number in standard form is written $a \times 10^n$ where n is an integer and a is a number between 1 and 10.

Example: Express 6530 in standard form.
$$6530 = 6.53 \times 1000 = 6.53 \times 10^3$$

Example: Express 0.0247 in standard form.
$$0.0247 = 2.47 \div 100 = 2.47 \times 10^{-2}$$

Express the following numbers in standard form:

1. 57.9
2. 34790
3. 624
4. 38.2
5. 970000
6. 3824000
7. 21.6
8. 9.64
9. 56800
10. 417.8
11. 0.539
12. 0.0684
13. 0.00295
14. 0.0102
15. 0.00009246
16. 0.178
17. 0.006482
18. 0.0198
19. 0.06285
20. 0.0000453

Exercise 35
Significant Figures: This indicates the degree of accuracy required in an example. Sequences of O's at the end of an integral number or the beginning of a decimal fraction are not counted. If the digit after the last significant figure required is 4 or less the last significant figure is unchanged. Similarly if the following digit is 6 or more increase the last significant figure by one. If the last significant figure is **odd** and is followed by a 5 increase the last significant figure by one and if the last significant figure is **even** followed by a 5 leave alone.

A similar method may be used when "rounding off" decimal examples to a prescribed number of decimal places.

Example: Express correct to a) 4 significant figures b) 3 significant figures 36.465.

Answer to a) 36.46

Answer to b) 36.5

Express correct to the prescribed number of significant figures.
1. 56.843 (3)   2. 0.079341 (4)
3. 7580. (2)   4. 3.9507 (4 and 2)
5. 0.0057964 (2)   6. 71.24 (3)
7. 418.52 (3 and 2)   8. 0.04049 (3 and 2)
9. 21.58 (2)   10. 29.564 (2)

Exercise 36

Express correct to the prescribed number of places of decimal.
1. 2.98364 (3 and 2)   2. 4.57426 (4 and 2)
3. 0.185253 (4)   4. 1.3257 (3)
5. 1.2046 (2)   6. 0.34528 (4 and 2)
7. 0.006738 (4)   8. 3.60812 (2)
9. 73.08361 (3 and 1)   10. 9.6163 (2)

CHAPTER 5. Square Roots. By inspection, by factors, using square root tables and iterative method.
Problems involving square roots.

EXERCISES 37 − 45

Exercise 37
Square Roots. When a number is multiplied by itself the product is called the square of the number; for example 6 x 6 = 36.
The reverse process is known as finding the square root of the number.
The symbol $\sqrt{}$ or $\sqrt{}$ is used to indicate a square root.
Example: $\sqrt{64}$ = 8.
Write down the square roots of:
1. 25   2. 49   3. 16   4. 81   5. 121
6. 10000   7. 196   8. 900   9. 169   10. 625

Exercise 38
Numbers which are perfect squares can often be broken up into factors and the square root found by inspection.

Example: Find $\sqrt{324} = \sqrt{2^2 \times 3^4} = 2 \times 3^2 = 18$

Find the prime factors of the following numbers and then their square roots.
1. 576   2. 441   3. 1225   4. 1764   5. 2916
6. 9216   7. 8649   8. 1089   9. 1444   10. 3481

Exercise 39
Find the exact square root of:
1. $\frac{25}{16}$ 2. $\frac{49}{64}$ 3. $\frac{16}{81}$ 4. $\frac{144}{25}$ 5. $2\frac{41}{64}$
6. $7\frac{1}{9}$ 7. $4\frac{21}{25}$ 8. $11\frac{1}{9}$ 9. $7\frac{9}{16}$ 10. $2\frac{14}{25}$

Exercise 40
Use square root tables to find to 3 significant figures the square roots of:
1. 2   2. 11   3. 29   4. 5.6   5. 17.2
6. 39.8   7. 84.3   8. 92.1   9. 40.6   10. 5.13

Exercise 41
If the number does not lie between 1 and 100 then its form must be changed to $K \times 10^m$ where K lies between 1 and 100 and m is an **even** power of 10.
Example: Find $\sqrt{5490}$ to 3 significant figures.
$$\sqrt{5490} = \sqrt{54.9 \times 100} = 7.41 \times 10 = \mathbf{74.1}$$
Find the square roots of the following numbers to 3 significant figures.
1. 839   2. 6720   3. 1990   4. 56300
5. 982   6. 702000   7. 1230   8. 810
9. 2250   10. 418000

Exercise 42
Example: Find $\sqrt{0.0692}$ to 3 significant figures.
$$\sqrt{0.0692} = \sqrt{\frac{6.92}{100}} = \frac{2.63}{10} = \mathbf{0.263}$$
Find the square roots of the following numbers to 3 significant figures.
1. 0.964   2. 0.0375   3. 0.0809   4. 0.000149
5. 0.614   6. 0.049   7. 0.0000863   8. 0.0532
9. 0.123   10. 0.00754

Exercise 43
Use square root tables in the following examples where necessary.
Find the square roots of:
1. $5\frac{4}{9}$ 2. 16900   3. 16.9   4. 2.25

5. 17.8   6. 0.0837   7. 59.2   8. 592
9. 0.000625   10. 91.6

Exercise 44

Find the square root of the following numbers by an iterative method.

a) 56.7   b) 0.0732

Make a rough estimate of the square root of the given number.

a) $\sqrt{56.7}$ = First estimate of 7.5

```
         7.56
    75 ) 567         Find average of 7.5 and 7.56
         525
         ---
         420
         375                        7.5
         ---                        7.56
         450              2 ) 15.06
         450                        7.53
         ---                        ====
```

$\sqrt{56.7}$ = **7.53** to 3 significant figures.

b) $\sqrt{0.0732} = \sqrt{\dfrac{7.32}{100}} = \dfrac{2.6}{10}$

2.6 is a first estimate.

```
         2.815
    26 ) 73.2              Find average of 2.6 and 2.815
         52
         ---
         212
         208                        2.815
         ---                        2.6
         40               2 ) 5.415
         26                         2.708
         ---                        =====
         140
         130
         ---
         10
         ==
```

$\sqrt{0.0732} = \dfrac{2.71}{10} = $ **0.271** to 3 significant figures.

If more significant figures are required a second division would have to be calculated using the last answer obtained (i.e. 7.53 or 0.271) as the divisor.

Find the square roots of the following numbers by an iterative method to three significant figures.

1. 14
2. 6.95
3. 87.2
4. 413
5. 0.0892
6. 0.0123
7. 0.00199
8. 71.4
9. 2.34
10. 11.6

Exercise 45

1. A ladder 14 metres long is placed against a vertical wall with its foot on horizontal ground and 5 metres from the wall. Find to one decimal place how high up the wall the ladder reaches.
2. A garden in the shape of a rectangle has one side of length 35 metres and its diagonal is 91 metres. Find the length of the remaining side.
3. A square field contains 22500 m^2. Find its perimeter and cost of fencing 3 of its sides at 23p per metre.
4. A square field contains 4225 m^2. Find the length of its sides and cost of sowing it with grass seed at 11p per m^2.
5. A rectangular field is twice as long as it is broad. If its area is 24200 metres^2 find its length and breadth.
6. One third of a number times two-fifths of the same number equals 480. What is the number?
7. A hall is five times as long as it is broad. If its area is 2880 m^2 find its length and breadth.
8. A carpet is 21 times as long as it is broad. If its area is 47¼ m^2 find its length and width. Find also its cost at £3.20 per m^2.
9. A school playground is 100 metres long and 75 metres broad. A drain runs diagonally across it. Find the length of piping required for this drain and cost at 16p per metre.
10. Find the length of the diagonal of a square of side 10 metres.
11. A rhombus has diagonals of length 6.3 cm and 8.4 cm. What is the length of its sides?

12. One third of a number times one quarter of the same number is equal to 5.88. What is the number?

13. A cyclist travels round a rectangular field whose length is four times its breadth. If the field contains 62500 m² and the cyclist is travelling at 40 Kilometres per hour find in seconds how long it takes him to travel round the field once.

14. Find the smallest whole number by which 7350 must be multiplied so that the result will have an exact square root.

15. 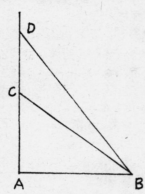 BC and BD are guy ropes attached to a flag pole.
BD is 20 metres long and AB is 12 metres. The distance between C and D is 7 metres; find the difference in lengths of the guy ropes.

CHAPTER 6. Sequences. Binary numbers. Different number bases.

EXERCISES 46 – 56

Exercise 46
Sequences or series: A succession of numbers obtained in accordance with some given law is called a **sequence or series**.
Example: Find the next two terms in the sequence 3, 8, 13, 18 – –
Required numbers are 23 and 28; since we add 5 to each term in turn.

Find the next two terms in the following sequences.

1. 3, 6, 9, – –
2. 7, 5, 3, – –
3. 29, 25, 21, – –
4. 7, 3, –1, – –
5. 5, 1, –3, – –
6. 5, 6.2, 7.4, – –
7. 5, 9, 13, – –
8. 6, 10, 14, 18, – –
9. 3a, 3a–2, 3a–4, – –
10. 42, 47, 52, – –
11. 9, 3, 1, – –
12. $2, 3, \frac{9}{2}, - -$
13. 40, 20, 10, 5, – –
14. 3, –9, 27, –81, – –
15. 63, –21, 7, – –
16. $\frac{1}{16}, \frac{1}{8}, \frac{1}{4}, - -$
17. 6, –12, 24, – –
18. 108, –54, 27, – –
19. 9, 6, 4, – –
20. $\frac{1}{3}, \frac{1}{2}, \frac{3}{4}, - -$

Exercise 47
Find the first four terms of the following sequences whose nth term is:

1. $4n + 3$
2. $3n + 2$
3. $2n + 5$
4. $\dfrac{1}{3n + 1}$
5. $15 - 2n$
6. $n(n + 1)$
7. $n^2 + 1$
8. $\dfrac{2n}{n + 1}$
9. $\dfrac{3n + 1}{2n - 1}$
10. $\dfrac{n^2}{2} + 1$
11. $\dfrac{n + 4}{n^2}$
12. $n^2 + n$

13. $\dfrac{1}{n^2}$　　　14. $2^n + 3$　　　15. n^n

Exercise 48

Establish the formula for finding the nth term of the following sequences:

1. 2, 3, 4, 5, ...
2. 3, 5, 7, 9, ...
3. 2 x 3, 3 x 5, 4 x 7, 5 x 9, ...
4. 5, 6, 7, 8, ...
5. 10, 9, 8, 7, ...
6. 5 x 10, 6 x 9, 7 x 8, 8 x 7, ...
7. 2×1^2, 3×2^2, 4×3^2, 5×4^2, ...
8. $1\frac{2}{3}$, $3\frac{1}{3}$, $6\frac{2}{3}$, ...
9. $\dfrac{1}{1.2}$, $\dfrac{1}{2.3}$, $\dfrac{1}{3.4}$, ...
10. 7, 5, 3, ...
11. 10, 7, 4, ...
12. 2, 7, 12, ...
13. 3, 4, $5\frac{1}{3}$, ...
14. 50, 20, 8, ...
15. 2, 5, 10, 17, ...

Exercise 49

Binary Numbers: These are numbers in which the base is **two** and the only symbols used are 1 and 0.

The place values are units, twos, fours, eights, etc. or units, 2^1, 2^2, 2^3, 2^4, ... etc.

Example:　Change 110101 from base two to a denary number (base ten).

$$110101 = 1 + 0 \text{ (twos)} + 1 \text{ (four)}$$
$$+ 0 \text{ (eights)} + 1 \text{ (sixteen)} + 1 \text{ (thirty-two)} = 53$$

Change the following binary numbers (base two) to denary numbers (base ten).

1. 1011
2. 1101
3. 10011
4. 110110
5. 10101
6. 11101
7. 1011101
8. 110111
9. 1101101
10. 11011101

Exercise 50

To change a number in base ten (denary) to a number in base two (binary) divide by successive twos until last quotient is a zero. The answer is read from the bottom upwards.

Example: Change 27 (base ten) to a binary number.

```
2 | 27
2 | 13   R1
2 |  6   R1
2 |  3   R0
2 |  1   R1
  |  0   R1
```

$27_{ten} = 11011_{two}$

Change the following base ten numbers to binary numbers.

1. 23
2. 41
3. 39
4. 57
5. 50
6. 67
7. 74
8. 93
9. 114
10. 85

Exercise 51

Addition Table for binary numbers.

+	0	1
0	0	1
1	1	10

Example: Add together the following numbers in base two.

10110 + 1011 = 100001

Find the sum of the following binary numbers:
1. 101 + 111
2. 1101 + 101
3. 1110 + 1011
4. 1101 + 111 + 11
5. 10111 + 110 + 11
6. 1110 + 1010 + 101
7. 1010 + 1101 + 111
8. 1011 + 1010 + 1101
9. 10101 + 1101 + 101
10. 10100 + 1111 + 1011

Exercise 52
Find the results of the following subtraction sums.
1. 10111 − 1101
2. 10101 − 1101
3. 11101 − 1011
4. 11111 − 10110
5. 100000 − 1111
6. 100100 − 11010
7. 1011001 − 110111
8. 101001 − 10100
9. 11011011 − 1011010
10. 1110110 − 11001

Exercise 53
Multiplication Table for binary numbers.

X	0	1
0	0	0
1	0	1

Example: Multiply together 1011 by 1101.

```
     1011
     1101
     ————
     1011
    10110
    1011
    ————
 10001111
```

Multiply together the following binary numbers.
1. 111 x 110
2. 101 x 11
3. 1101 x 101
4. 1011 x 110
5. 11011 x 1101
6. 10101 x 1011

7. 110111 x 111   8. 11010 x 1011
9. 11110 x 110   10. 100001 x 1010

Exercise 54
Division of binary numbers.

Example: Find the quotient and remainder in the following division
10110 ÷ 101

```
          1001
     ┌─────────
 101 │ 101110
       101
       ───
        110
        101
        ───
          1
```

Quotient = 1001
Remainder = 1

Find the quotients and remainders in the following divisions.

1. 10011 ÷ 11   2. 110111 ÷ 101
3. 110110 ÷ 110   4. 111011 ÷ 111
5. 1010101 ÷ 1101   6. 10111011 ÷ 111
7. 111111 ÷ 1010   8. 1000000 ÷ 1100
9. 10100100 ÷ 1100   10. 1101100 ÷ 1111

Exercise 55
Conversion to and from bases other than two can be done in a similar fashion as that for binary numbers.

For base 5 the place values are units, fives (5') twenty-fives etc.

For base three the place values are units threes (3'), nines (3^2), twenty-sevens (3^3) etc.

Convert the following numbers from the bases indicated to denary numbers (base ten).

Example: a) 2102 three = 2 x 27 + 1 x 9 + 0 x 3 + 2 = 65.
           b) 3201 four = 3 x 64 + 2 x 16 + 0 x 4 + 1 = 225.

1. 1210 three
2. 12110 three
3. 2201 three
4. 123 four
5. 320 four
6. 3321 four
7. 3012 four
8. 123 five
9. 4301 five
10. 1431 five
11. 4430 five
12. 63 eight
13. 126 eight
14. 147 eight
15. 265 eight

Exercise 56

To convert from base ten to a given base — for example 5 — divide by successive fives until last quotient is zero.

Example: Change 297 base ten to base five.

```
5 | 297
5 |  59   R2
5 |  11   R4
5 |   2   R1
        0  R2
```

297 ten = 2142 five

Change the following base ten numbers to numbers in the bases indicated.

1. 39 (three)
2. 57 (four)
3. 69 (five)
4. 81 (eight)
5. 72 (three)
6. 105 (four)
7. 234 (eight)
8. 195 (five)
9. 61 (three)
10. 270 (four)
11. 93 (eight)
12. 123 (five)
13. 74 (five)
14. 326 (eight)
15. 394 (five)

CHAPTER 7. The metric system. Conversions. Area, volume, capacity weight. Problems on metric units.

EXERCISES 57 — 61

Exercise 57

The **metric system** derives its name from the **metre** which is the unit of length.

The metre is sub-divided into tenths, hundredths and thousandths and the Latin prefixes for these are deci, centi and milli.

The multiples are Greek prefixes Deko, Hecto and Kilo which are ten times, one hundred times and one thousand times, respectively:

    10 Millimetres = 1 Centimetre
    10 Centimetres = 1 Decimetre
    10 Decimetres   = 1 Metre
    10 Metres   = 1 Dekametre
    10 Dekametres   = 1 Hectometre
    10 Hectometres = 1 Kilometre

A metre is slightly more than 1 yard in length and a Kilometre is approximately 5/8 mile.

Example: Reduce   5 m   6 dm   8 cm to cm
                                = 568. cm

Example: Change   0.5893 km to m
                           0.5893 km = 589.3 m

The **reduction** from one denomination to another in the metric system is effected by altering the position of the decimal point.

Reduce:
1. 8 m 5 cm to m and cm
2. 5 m 6 dm to cm and mm
3. 8 Km 56 m to Km and m
4. 8 Hm 5 Dm 96 cm to m and Dm

5. 6 m 3 cm 8 mm to cm and mm
6. 4625 m to Km
7. 3496 cm to m
8. 14936 cm to m and Km
9. 6947 m to Dm and Km
10. 0.6327 Km to m and cm

Exercise 58

The **unit of area** is the **square metre** and to reduce from one denomination to the next lower one we multiply by 100.

Example: Change $4\,Km^2$, $63\,Hm^2$, $9\,Dm^2$, $64\,m^2$ to m^2.
$$= 4 \times (1000)^2 + 63 \times (100)^2 + 9\,(10)^2 + 64$$
$$= 4630964\,m^2.$$

Change:
1. $3\,m^2$, $27\,cm^2$ to cm^2
2. $9\,Km^2$ to Dm^2
3. $1\,Hm^2$, $56\,m^2$ to m^2
4. $7\,m^2$, $64\,dm^2$ to cm^2
5. $6\,Km^2$, $493\,Dm^2$ to Dm^2
6. $2974\,m^2$ to Dm^2
7. $57932\,cm^2$ to m^2
8. $7659302\,mm^2$ to m^2
9. If an are = 1 decametre^2 and 1 Hectare = 100 ares find the area of a field in Hectares which is rectangular in shape and is 147 m long and 108 m broad.
10. Change $579362\,m^2$ to ares and Hectares.

Exercise 59

The **unit of volume** is the **cubic metre** and to reduce from one denomination to the next lower one we multiply by 1000.

Example: Change $5\,m^3$ $693\,cm^3$ to cubic centimetres

$$= 5 \times (100)^3 + 693$$
$$= 5000693 \text{ cm}^3$$

The **unit of capacity** is the litre (l) and 1 litre = 1000 cubic centimetres.
Reduce:
1. 69 m^3, 465 dm^3 to dm^3
2. 798374 cm^3 to m^3 (answers to 2 sig. figs.)
3. 6 cm^3, 974 mm^3 to cm^3
4. 8436 m^3 to Dm^3
5. 98537 cm^3 to litres
6. 9 Hl to litres
7. 7685 ml to cl and l
8. 53 Kl, 2 Hl to l
9. 16843 cl to l and Hl
10. 0.06943 Kl to l

Exercise 60
The **metric unit** of **weight** is the **gramme (g)**.
Reduce:
1. 56 Kg, 3 Hg to Dg and g
2. 34789 cg to g and Hg
3. 93 Hg, 47 g to g
4. 5 Kg, 3 Hg, 6 Dg, 9 g to Hg and g
5. 47 Kg to cg
6. 8937618 cg to g and Kg. (Answer to one decimal place).
7. 96 g, 47 cg to Dg
8. 375834 g to Kg and cg
9. 5 Kg to mg
10. 12964538 mg to g and Kg. (Answer to 2 decimal places).

Exercise 61
1. A petrol storage tank is rectangular in shape and measures 3 m by 2 m by 1.5 m. Find its capacity in litres.

If the garage sells 500 litres of petrol per day, how many days supply does the tank hold?

2. A tank contains 43.6 Kl. If 7.3 Hl must be added to fill the tank find the total capacity of the tank in litres.
3. A bottle contains 5 litres of liquid. How many containers of capacity 75 cm^3 can be filled from it and what volume is left?
4. A grocer orders 14 quintals of sugar. How many customers can be supplied with 4 Kg each if 1 quintal equals 100 Kg?
5. A coal merchant orders 23 truck loads of coal per week. Each truck contains 16 tonnes and the coal is put into bags holding 1 quintal. Find the number of bags of coal the merchant uses per week.
(1 tonne = 1000 Kg and 1 quintal = 100 Kg).
6. If a metre of wire weighs 6.2 g find the weight of a coil of wire of length 76.5 m.
7. A swimming pond is 25 metres long, 2 metres deep and 10 metres wide.
How long will it take to fill it if water is pouring in at the rate of 5 Kilolitres per minute?
8. A father is 26.9 cm taller than his son. If the son is 1.47 metres tall what is the father's height in mm.
9. A supply tank contains 6.3 Kl. If 1743 litres are drawn off how many Dl are left?
10. A classroom is 8 metres long, 7 metres wide and 3 metres high. If it accommodates 40 pupils what volume of air is each pupil being allowed?

CHAPTER 8. Direct Proportion. Inverse Proportion. Averages and Speed Averages.

EXERCISES 62 − 66

Exercise 62
Direct Proportion

Example: If 5 articles cost £1.30 find the cost of 12 at the same rate.

$$5 \text{ articles} \longleftrightarrow £1.30$$
$$12 \text{ articles} \longleftrightarrow £1.30 \times \frac{12}{5} = £3.12$$

1. If 21 articles cost £7.35 find the cost of 27 at the same rate.
2. A car can run 154 Km on 14 litres of petrol. How far will it run on 23 litres?
3. For working 44 hours a man earns £39.60. How much should he be paid for working 52 hours at the same rate?
4. In a house where the assessed value is £108 a man pays £135 in rates per year. Another house in the same district has an assessed value of £96. Find the difference in the amounts paid in rates.
5. On a map the length of a certain rectangular area is 4.5 cm. If the area actually measures 24 Km by 16 Km what is its breadth on the map?
6. A cyclist travels 55 Km in 2 hours 45 minutes. How long will he take on a journey of 100 Km?
7. If the rent of 144 hectares of land is £594 per year, how many hectares can be rented for £990?
8. If I can read 27 pages in 45 minutes how long will it take me to read a book containing 320 pages?
9. If 5 brick layers can lay 2650 bricks in a day, how many brick layers will lay 5830 bricks in a day?
10. The employees in a certain firm were given a sum proportional to their length of service.
    A man who had served 7 years received £131.25. How long had an employee served who received £405?

Exercise 63
Inverse Proportion

Example: If 30 men take 24 days to do a piece of work how long would 16 men have taken?

$$30 \text{ men} \longleftrightarrow 24 \text{ days}$$
$$16 \text{ men} \longleftrightarrow \frac{\cancel{24}^{3} \times \cancel{30}^{15}}{\cancel{16}_{1} \cancel{2}_{1}} = 45 \text{ days}$$

1. If 500 cattle have enough fodder for 75 days, how long should the fodder last 625 cattle?
2. Fifteen men take 12 days to build a new road. How many men will be employed if the road is built in 20 days?
3. A ship has enough oil in its tanks to last it 12 days if it uses 4.2 Kl per day. If she uses 5.6 Kl per day, how long should the oil last.
4. How many metres of cloth worth 72p per metre should be given for 40 metres at 96p per metre?
5. A garrison of 1500 men have provisions for 48 days. How long will the food last if 300 men leave the garrison?
6. A train completes a journey in 15 hours travelling at 56 Km/hr. At what rate per hour must it travel to complete the journey in 12 hours?
7. 1350 men have sufficient food to last 12 months. How long will the food last if 450 extra men arrive?
8. A piece of work is completed by 63 men in 32 days. How many extra men are required to complete the work in 28 days?
9. A set of book shelves contains 360 books of an average width of 4½ cm. How many books of average width 3¾ cm will it hold?
10. How long would an aeroplane travelling at 720 Km/hr take for a certain journey if it takes 6 hours travelling at 900 Km/hr?

Exercise 64

Averages: The **average** of a number of quantities is their sum divided by the number of quantities.

Example: Find the average of 18, 16, 22 and 20.

$$\text{Average} = \frac{18 + 16 + 22 + 20}{4} = 19$$

1. Find the average of 111, 106, 94 and 85.
2. Find the average of 23.6, 16, 13.1, 6.5 and 5.3.
3. Find the average of 13.0, 9.4, 7.8, 5.0, 4.6, and 3.4.
4. The ages of 6 boys are 9, 10, 13, 14, 15 and 17. Find their average age.
5. A batsman scored 45, 51, 71, 0, 105, 13 and 93 runs in successive cricket innings. Find his average number of runs per innings.
6. A cyclist takes 2 minutes 5 seconds, 2 minutes 25 seconds, 2 minutes 35 seconds, 2 minutes 15 seconds and 2 minutes for five successive 1 Km sprints. Find his average time per Km and his average speed in Km/hr.
7. The rainfall during six consecutive months in a certain district was 5.54 cm, 8.08 cm, 22.33 cm, 28.94 cm, 13.63 cm and 5.27 cm. Find the average rainfall per month for this district to 2 decimal places.
8. The takings in a shop for a week of 6 days were £147.36, £84.22, £150.85, £136.91, £208.24 and £235.73. Find the average daily takings to nearest pence.
9. The heights of 6 policemen were 173.6 cm, 184.2 cm, 194.5 cm, 188.3 cm, 179.5 cm and 191.7 cm. Find the average height to one decimal place and by how much the smallest policeman is below this average.
10. In a school of 1200 pupils the number of absentees on 10 consecutive openings were 96, 125, 87, 94, 108, 133, 154, 142, 113 and 118. Find the average number present per opening.

Exercise 65

1. The average weight of a boat crew and coxwain is 86 Kg. The average weight of the eight oarsmen is 87.75 Kg. Find the weight of the coxwain.
2. The average attendance at a school concert running for 4 nights

was 730. The attendances on the first 3 nights were 685, 742 and 717. Find the attendance on the fourth night.

3. There are 38 pupils in a class and their average mark for a French examination was 58. If the average of the 16 boys was 53 find the average mark of the girls to one decimal place.

4. A man employs 140 men. Three are paid £45, 12 are paid £35, 60 are paid £30 and the rest are paid £22 per week. Find the average wage paid per week to nearest pound.

5. The average age of 12 men is 43. Three more join the group, two of whose ages are 37 and 41. Find the age of the last man if the average of the whole group remains at 43.

6. A small business has an average profit per month over a 6 month period of £230. During the first 3 months it made average losses of £126 per month and during the fourth and fifth months profits of £87 and £274. Find the profit during the sixth month.

7. Find the average cost per book of 10 books, 6 of which cost 30p each and the rest cost 50p each.

8. In a shooting team of six the best marksman scored 90 points. If he had scored 96 points the average for the team would have been 87. Find the total actually scored by the team.

9. A cyclist intends to cover 500 Km in 7 days. For the first 5 days he averages 67 Km per day and on the sixth day he cycles 84 Km. How many kilometres must he cycle on the last day to complete the journey?

10. During seven weeks a man earned £31.43, £36.82, £45.26, £32.18 £46.52, £40.27 and £38.16. If during the next five weeks he earned enough to make his average, during the twelve week period, £42.25 per week, find his average weekly earnings during the last five weeks to nearest pence.

Exercise 66

Speed Averages: If a journey consists of several parts done at different speeds find the time taken for each.

The average speed = $\dfrac{\text{Total Distance}}{\text{Total Time}}$

1. A train completes a journey of 670 Km in 9 hours. During the first 3½ hours it travels at an average speed of 64 Km/hr and during the next 4 hours its average speed is 78 Km per hour. Find its average speed during rest of journey.
2. A motor car completes a journey of 560 Km in 9 hours 25 minutes. The driver has 3 stops on the way. One of 15 minutes and two of 20 minutes. Find the average speed of the car while in motion. (Answer to 1 decimal place).
3. A cyclist set out at 9 a.m. on a journey of 165 Km. He stopped for 45 minutes for lunch and arrived at his destination at 4.15 p.m. Find his average speed while in motion (answer to 1 decimal place).
4. A man travelled 60 Km at 60 Km/hr and the next 60 Km at 90 Km/hr. Find his average speed for the whole 120 Km.
5. If I travelled for 1½ hours at 48 Km/hr and for 2 hours at 64 Km/hr find my average speed in Km/hr to the nearest whole number.
6. A motorist covered 344 Km at an average speed of 43 Km/hr. He travelled at 50 Km per hour for the first 80 Km and during the next 140 Km he travelled at 40 Km per hour. Find his average speed during the last part of the journey in Km/hr. (Answer to the nearest whole number).
7. On a journey of 160 Km a motorist covered the first 64 Km at a speed of 80 Km per hour. He then travelled at 48 Km per hour for the next half-hour. If the journey took 2 hours 28 minutes altogether, what was his speed in Km/hr for the last part of the journey? (Answer to 1 decimal place).
8. A traveller intends covering 432 Km by car in 9 hours. He does the first half of the journey at 54 Km per hour and is then delayed with engine trouble for 1 hour 15 minutes. At what speed per hour must he complete the journey if he wishes to arrive at the time intended? (Answer to 1 decimal place).
9. The 9.30 a.m. train from Glasgow to Inverness, 268 Km away, arrives at 3.20 p.m. If the train had a 15 minute stop at Perth find its average running speed in Km per hour.
10. Two motorists start from the same point at 10.30 a.m. and proceed in opposite directions until 12.15 p.m. when the distance

between them is 196 Km. If one motorist had been travelling at 48 Km/hr what must have been the speed of the second one?

CHAPTER 9. Profit and loss per cent. Finding selling prices and cost prices. Simple and Compound Interest. Compound increases and decreases. Percentage discounts and commissions.

EXERCISES 67 — 80

Exercise 67
Profit and Loss: A shopkeeper makes a **profit** if he sells an article for **more** than it cost him. He makes a **loss** if he sells it for **less** than it cost him.

1. A grocer buys 100 Kg of tea at 56p per Kg and sells it at 18p for ¼ Kg. Find his total profit.
2. The cost of producing 10,000 copies of a hit record is £3000. The records are sold at 96p each and after paying the production cost the company pays a tax of 40% and 15% to the composer of the hit. Find what profit remains for the company.
3. A length of cloth containing 100 m cost a shopkeeper £69. He sold 87 m of it at 92p per metre and the rest during a sale at 64p/m. Find his total profit.
4. A fruiterer bought 10 cases of apples each containing 20 Kg for a total of £43. He sold them all at 17½p per Kg. Find profit or loss.
5. A butcher bought 5 sides of beef each weighing an average of 112 Kg for a total of £300. He sold 420 Kg of the beef at 84p per Kg, a further 80 Kg at 69p per Kg and the rest at 42p per Kg. Find his profit.
6. A horse Breeder bought 8 horses for £736. He later sold 3 of them for £107 each, 4 of them for £85 each and the last one for £72. Find his loss on the deal.
7. A stationer buys 6 packets of notebooks each containing 100 notebooks for a total of £29. He sells 4½ packets at 8p per notebook and the rest at 6p per notebook. Find his total profit.

8. A householder pays a deposit of £39 for a television set and then 12 instalments of £5.60. Later he is forced to sell the set for £73. Find his loss.
9. A coal merchant receives a load of 6 tonnes of coal which costs him £116.40. He sells it all at £2.56 per quintal. Find his profit or loss. (1 tonne = 1000 Kg. 1 quintal = 100 Kg).
10. A market gardener planted out 10,000 lettuces. The cost of the seed was £3, fertiliser £16 and labour £230. During the season he sold 8,500 of them at an average of 5½p each but the rest had to be thrown away. Find his profit or loss on the transaction.

Exercise 68
Profit or Loss Per Cent: We express the profit or loss as a fraction of the cost price and then convert this fraction to a percentage.

Example: A man bought a house for £5280 and later sold it for £5676. Find his profit per cent.

$$\text{Profit} = £5676 - £5280 = £396$$

$$\text{Profit per cent} = \frac{396}{5280} \times \frac{100}{1} = \frac{15}{2}\% = 7\tfrac{1}{2}\%$$

Find the profit or loss per cent in the following examples:
1. Cost price £10.00 : selling price £12.50.
2. Cost price £11.25 : selling price £13.50.
3. Cost price £ 8.40 : selling price £ 5.25.
4. Cost price £12.00 : selling price £15.00.
5. Cost price £ 3.30 : selling price £ 3.96.
6. Cost price £26.25 : selling price £30.00.
7. Cost price £ 2.40 : selling price £ 1.80.
8. Cost price £40.00 : selling price £33.00.
9. Cost price £82.50 : selling price £75.00.
10. Cost price £ 5.25 : selling price £ 4.50.

Exercise 69

1. A piano which cost £120 was sold for £145. Find the gain per cent
2. A farmer bought a cow for £90 and later sold it for £105. Find his gain per cent.
3. Curtain material which cost 75p per metre was sold at 92p per metre. Find the gain per cent.
4. A house which cost £2950 in 1960 was sold in 1973 for £12500. Find the profit per cent to nearest whole number.
5. A bookseller bought 800 books at 75p each. He sold 710 of them at £1.06 each and the rest at half their normal selling price. Find his gain and gain per cent. (Answer to 1 decimal place).
6. A car which cost £1200 when new was sold two years later for £935. Find the loss per cent.
7. A coal merchant paid £18.75 per tonne of coalite. He sold it at £2.50 per quintal. Find his gain or loss per cent.
   (1 tonne = 1000 Kg.   1 quintal = 100 Kg)
8. A cycle dealer sold a cycle for £37.50. Find his gain per cent if it cost him £30.
9. A stationer bought Christmas cards at £1.75 per hundred and he sold them in packets of ten costing 24p each. Find his profit per cent.
10. A fruiterer bought 30 boxes of oranges each containing an average of 100 oranges at £2 per box. He also had to pay £2.50 for transport charges.
    The oranges were marked for sale at 3 for 10p and 24 boxes were sold at this price. The remainder were sold at 5 for 10p. Find the fruiterers profit or loss per cent. (Answer to 1 decimal place).

Exercise 70

Finding Selling Price: The cost price of an article is regarded as being 100% and the selling price will be more or less than this depending on whether a gain or loss per cent is made.

Example: An article which cost £15 is sold at a profit of 20%. Find the selling price.

$$100\% \longleftrightarrow £15$$
$$120\% \longleftrightarrow \frac{\cancel{15}^{\,3}}{1} \times \frac{\cancel{120}^{\,6}}{\cancel{100}_{\,2}} = £18$$

Find the selling price in each case.

1. Cost price £15.60 : loss 30%
2. Cost price £22.00 : profit 15%
3. Cost price 42p : profit $33\frac{1}{3}$%
4. Cost price £8.40 : loss $8\frac{1}{3}$%
5. Cost price £25.00 : loss 20%
6. Cost price £750 : profit 40%
7. Cost price £2550 : profit $16\frac{2}{3}$%
8. Cost price £2.76 : loss 25%
9. Cost price £63 : profit 30%
10. Cost price £1.40 : loss 2½%

Exercise 71

Finding the Cost Price: These examples are done in a method similar to the previous exercise, that is by simple proportion.

Example: Find the cost price of an article which was sold for £10.80 at a loss of 10%.

$$\text{Selling price} = \text{Cost price} - \text{Loss}$$
$$= 90\% \text{ of cost price}$$
$$90\% \longleftrightarrow £10.80$$
$$100\% \longleftrightarrow £10.80 \times \frac{100}{\cancel{90}} = \frac{108}{9} = £12$$
$$\text{Cost Price} = £12$$

Find the cost price in each case.

1. Selling price £4.86 : profit 12½%

2. Selling price £13.60 : loss 15%
3. Selling price 84p : loss $33\frac{1}{3}$%
4. Selling price 10½p
   per Kg : profit $16\frac{2}{3}$%
   Find the cost price per quintal (100 Kg)
5. Selling price £14.50 : profit 45%
6. Selling price £2.20 : profit 37½%
7. Selling price £4.98 : loss 17%
8. Selling price £2160 : profit 25%
9. Selling price £22.50 : loss $16\frac{2}{3}$%
10. Selling price £1.95 : loss 2½%

Exercise 72
Mixed Examples

1. By selling a house for £7560 the owner gained 12%. Find how much the house cost him.
2. A fruiterer bought 3000 bananas at 16p per doz. He sold 1800 of them at 3p each and the rest at 3 for 8p. Find his gain per cent.
3. A stereo unit was priced at £84 but the shopkeeper agreed to reduce it by 12½% since it was shop soiled. At this price he still made a profit of 5%. Find the cost price of the stereo unit.
4. If I increase the price of an article by 68p a loss of $6\frac{2}{3}$% is turned into a gain of 10%. Find the cost price of the article.
5. A second-hand car cost me £750. Two months later I had to sell it at a loss of 12½%. For how much did I sell it?
6. A cycle dealer sold a bicycle for £26.91 and made a profit of 30%. Find how much the bicycle cost him.
7. When an article is sold for £11.00 a profit of 10% is made. Find the percentage gain or loss if the article is sold for £9.50.
8. A dealer bought a horse for £157.50 and later sold it at a gain of $16\frac{2}{3}$%. For what price did he sell it?
9. A butcher sold a side of lamb at a loss of 5%. If he had charged

£1.65 more for it he would have gained 20%. Find the cost of the side of lamb.

10. A shopkeeper bought 144 metres of cloth for £57.60, which he sold at 70p per metre. Find his gain or loss per cent.

Exercise 73

Simple Interest: When a sum of money is deposited (principal) in a bank the bank makes use of this money for further investments. The depositor is rewarded by the bank for the use of his money by the payment of interest. This is normally calculated on a yearly basis and depends on the rate per cent given by the bank.

Example: Find the simple interest on £650 for 1 year at 5% per annum.

$$\text{Interest} = 5\% \text{ of } £650 = \frac{1}{20} \times \frac{£650}{1} = £32.50$$

If we add the principal and the interest £650 + £32.50 = £682.50 we obtain the **amount**.

Find the simple interest for **one** year on the following sums of money at the rates given.

1. £390 invested at 3%.
2. £440 invested at 4%.
3. 1600 dollars invested at 6%.
4. 280 francs invested at 2½%.
5. £720 invested at 5%.
6. £15 invested at 9%.
7. 6000 dollars invested at 3½%.
8. £560 invested at 3¾%.
9. 450 francs invested at 6¼%.
10. 720 francs invested at 10%.

Exercise 74

Find the amount after one year of the following sums of money at the rates given:
1. £480 invested at $4\frac{1}{3}$%.
2. £560 invested at 3½%.
3. 220 dollars invested at 2¼%.
4. 780 francs invested at 6%.
5. £5400 invested at $8\frac{1}{6}$%.

Exercise 75

If the principal is only invested for a few months the interest will be the fractional part of the year counting only complete months.

Find the simple interest on the following sums of money. (Answers to the nearest pence).
1. £510 invested for 8 months at 2% per annum.
2. £840 invested for 10 months at 4% per annum.
3. 3500 dollars invested for 6 months at 4½% per annum.
4. 1080 francs invested for 8 months at 3¾% per annum.
5. £184 invested for 10 months at 6¼% per annum.
6. 1120 dollars invested for 6 months at 7% per annum.
7. £250 invested for 9 months at 7½% per annum.
8. £798 invested at 5% per annum for 7 months.
9. £467 invested at 6¼% per annum for 10 months.
10. £837 invested at $6\frac{2}{3}$% per annum for 4 months.

Exercise 76

Compound Interest: In compound interest the interest is added to the Principal at the end of each year and this amount forms the principal for the following year.

Calculations are normally done only on complete pounds.

Example: Find the interest on £560 for 3 years at 5% compound interest.

Principal for 1st year	=	£560	5.60
Interest for 1st year = 5% of £560	=	£28	5
			28.00
Principal for 2nd year	=	£588	5.88
Interest for 2nd year = 5% of £588	=	£29.40	5
			29.40
Principal for 3rd year	=	£617.40	6.17
Interest for 3rd year = 5% of £617	=	£30.85	5
			30.85
Amount for 3 years	=	£648.25	
Original Principal	=	£560	
Interest for 3 years	=	£88.25	

Find to the nearest £0.01 the compound interest on:
1. £620 for 2 years at 4%.
2. £790 for 3 years at 3%.
3. £400 for 3 years at 2½%.
4. £356 for 2 years at 5%.
5. £1500 for 3 years at 4%.
6. £380.64 for 2 years at 3½%.
7. £572.80 for 2½ years at 3%.
8. £960.36 for 2 years at 6½%.
9. £825.73 for 3 years at 5%.
10. £673.60 for 2 years at 8½%.

Exercise 77
Find the amount at compound interest to the nearest £0.01.
1. £650 in 3 years at 4%.
2. £270 in 2 years at 5%.
3. £1840 in 2½ years at 6%.
4. £564 in 3 years at 3%.
5. £620.85 in 2 years at 4½%.
6. £790.26 in 3 years at 7%.
7. £570.80 in 2 years at $3\frac{1}{3}$%.
8. £531.20 in 3 years at 5%.
9. £400 in 3 years at 3¾%.
10. £654.90 in 2 years at 6%.

Exercise 78
Compound increases and decreases: These examples are calculated in a similar manner to compound interest. For depreciation examples the percentage results are subtracted instead of being added.
1. The population of a new town is 25,000 and it is increasing at the rate of 10% per year. Find the population to the nearest whole number after three years.
2. A car cost £1520 when new and its value depreciates at the rate of 15% per year. Find its market value after 4 years to the nearest £.
3. A second-hand car is valued at £720 and it is reckoned its value will depreciate at the rate of 25% per year. Find how much it will be worth 3 years later.
4. The population of a large city was 950,000 at the beginning of 1970 and it is calculated that it is decreasing at the rate of 5% per year due to rehousing in new towns. Estimate to the nearest hundred the population expected at the end of 1973.
5. The cost of maintaining machinery in an engineering firm was £300 but due to increases in cost of parts and labour this was estimated to increase by 10% each year. Find the cost of mainten-

ance after 5 years. (Answer to nearest £).

6. The machinery in a firm cost £25,000 and it depreciates at the rate of 12½% per year. Find the value of the machinery after 3 years.
7. A small factory produces 78,000 plastic toys each month. Due to an incentive bonus the production each month is increased by 5% of the previous months total. Find the number of toys produced at the end of 4 months.
8. A house is valued at £8,500 and it is reckoned its value will increase by 10% each year, of its value, over a three year period. Find its value after the three year period to the nearest £.
9. A coal mine which is being run down produces 20% less each year than the previous one. During one year the production fell to 10,000 tonnes and it was decided to close down after a further 3 years. How much coal was produced during these last 3 years?
10. A television aerial consists of 6 parts, each of which weighs 90% of the part below it. What is the total weight of the aerial to the nearest Kg if the bottom section weighs 40 Kg? (Take percentage of whole numbers only).

Exercise 79

Percentage Discounts: This is a small amount by which the cost to a customer of an article is reduced during a sale or for ready cash.

In the following examples take answers to nearest pence.

1. A piano which is marked at £128 is offered for sale at a discount of 5%. How much cash is actually paid for the piano?
2. In a warehouse, discounts of 12½% are given to employees if payment for articles is made within one month of date of purchase and 7½% if payment is made between one and three months of purchase.

    Employee A bought a sink unit costing £65 and clothes costing £39.50. How much did he pay for these items if he paid within one month?

    Employee B paid two months later for items costing £76.30. How much did he actually pay?

3. During a sale an electricity board offered cookers at a discount of 5% after a first reduction of £5 on the marked price. Find the cost to a customer of a cooker marked at £88.
4. During a sale in a large store a discount of 10% on all items is given for ready cash. Find the amount paid for articles marked at a) £5.40; b) £16.20; c) £46.75.
5. During a sale shop soiled articles are sold at a discount of 15%. Find the price paid for:
   a) A pair of sheets marked at £3.75.
   b) A bedspread marked at £6.40.
   c) Bath towels which normally cost £1.64.
6. To make room for new stock a car salesman sold the previous years cars at a discount of 8%. Find the sum paid for a Vauxhall Victor marked at £1400.
7. A deep freeze company is offering for sale a deep freeze at £86 which normally cost £120. What percentage discount was given?
8. After discount a customer paid £2.40 for a shirt which normally cost £3.20. What was the percentage discount on the shirt?
9. A warehouse offered for sale a washing machine at £131.50. The normal price of the machine was £145. What discount was the warehouse offering?
10. A three piece suite which normally cost £246 was offered for sale with a discount of 8% for ready cash. Find the cash price. During an end of season sale a further 5% reduction was given on the reduced price. What is the reduced price during the sale? (Answer to nearest pence).

Exercise 80

Commission: This is a payment which is normally a fixed percentage of sales, given to salemen and is part of their salary.
1. An estate agent charges 1¼% for selling houses. Find his net profit for the sale of a house at £12,600 if he has expenses of a) £23 legal expenses; b) £1.76 telephone calls; c) £3.20 advertising.

2. Find an estate agent's net profit if he charges 1½% commission for selling a house for £9,670 and he incurs expenses of £15.84.
3. A store assistant is given commission of 2½% on all sales made by her calculated on whole pounds at the end of each week. What would be her commission during the week in which her daily sales were Monday £46.32; Tuesday £18.59; Wednesday £65.71; Thursday £60.87; Friday £89.36 and Saturday £102.16?
4. The assistants in a large store are given commission of 2% on their total sales and the manager of the department is given commission of 1¼% of the total of his staffs' sales. Find the commissions received by the following individuals.
   Assistant A — sales for week £362.47.
   Assistant B — sales for week £284.93.
   Assistant C — sales for week £509.81.
   Assistant D — sales for week £481.16.
   Manager   — (commission on whole pounds).
5. A car salesman is given a commission of 2% on the sale of a car up to £1,000 and 1¼% on the sale of a car over this figure. Find the commission he receives in the week in which he sells two cars at £870; 3 at £940; 1 at £1500 and 1 at £2350.
6. Find a car salesman's average weekly wage over a four week period if he is given a fixed sum of £8 plus commission of 1½% on the value of all his sales. In week 1 his sales amount to £5640, in week 2 they are £4275, in week 3 they are £6090 and week 4 £3956.
7. An insurance agent is given a fixed wage of £16 per week and 4% on the value of insurance he sells. Taking a year as 52 weeks, find his gross salary for the year in which he sells insurance valued at £27,000.
8. Find the total commission received by an insurance agent if he receives 4% for sums over £10,000 and less than £20,000 and 2½% on sums over £20,000. Find his commission in the year in which he sells £59,600 worth of insurance.
9. A house valued at £14,650 is sold by an estate agent who charges 1% commission. Find his net profit if he incurs expenses of £37.93.
10. An agent is given a fixed wage of £10 per week plus commission

of 3% on all sales over £15,000. Find his average weekly wage during the year in which his total sales amount to £73,600.

CHAPTER 10. Insurance, Assurance, Bankruptcy and Foreign Exchange and Travel.

EXERCISES 81 – 86

Exercise 81
Insurance is financial protection against certain types of risks; accident, fire, theft, death. The money paid to the insurance company to provide this protection is called the **premium** and it is usually a fixed rate per cent.
1. Calculate the premium on £3,500 at 1¼%.
2. Calculate the premium on £600 at $1\frac{1}{8}$%.
3. Calculate the premium on £15,700 at ½%.
4. Calculate the premium on £9,560 at 2%.
5. Calculate the premium on £830 at $1\frac{3}{8}$%.
6. Calculate the premium on £1,760 at 1¼%.
7. Calculate the premium on £5,630 at 1½%.
8. Calculate the premium on £19,000 at 2%.
9. Calculate the premium on £520 at $1\frac{1}{8}$%.
10. Calculate the premium on £6,070 at 2¼%.

Exercise 82
1. A householder insures the contents of his house valued at £950 at a premium of 30p per £100 per year. What is his annual premium?
2. A house is valued at £8470 and the householder pays an insurance policy on it for full coverage at a rate of £0.36 per cent. Find his annual premium.
   The value of his house increases to £11,500 and the householder increases his insurance coverage to meet the new value. What

increase will he have to pay in insurance?

3. The full premium to be paid by a car owner for a comprehensive policy on his car is £106. Since he is the sole driver he is given a 10% deduction on this figure and since he has driven for several years without accident this reduced figure is further reduced by 50%. Find the premium he actually pays.

4. A car owner takes out a third party only policy on his car which is second-hand. The full premium is £52.50 but he is given an initial deduction of 10% being the sole driver and this figure is then reduced by $33\frac{1}{3}\%$ for careful driving. What premium does he actually pay per year?

5. A ships cargo is valued at £63,000. What premium must be paid by the owners to the insurance company at 2½%?

6. The contents of a warehouse are insured at a rate of 3¼%. If the contents are worth £1,647,000 find the annual premium paid.

7. A fishing vessel is valued at £46,000 and it is insured for its full value at £2.70 per cent. What fraction of the ship does a shareholder own if he pays £138 worth of the insurance premium?

8. A property is insured for 7/8 of its value at $1\frac{3}{8}\%$. If the premium amounts to £154 per year what is the value of the property?

9. A woman insures her jewellery at a rate of 5%. Find the premium paid if her jewellery is worth £2,750.

10. A man insures his watch which was valued at £76 at a premium rate of 4½%. Find what he pays to insure his watch over a five year period.

Exercise 83

Life Assurance: This is a sum of money paid on a persons life and may be for a fixed period or whole life. In the case of the former, the person is paid a guaranteed sum after a certain number of years or to his dependents in the case of his death. In the latter type of policy the sum assured is only paid to dependents when the policy holder dies. On these types of policies a bonus is normally paid each year in the

same way as interest is paid in banks.
1. A man of twenty-five takes out a life assurance policy of £2,000 which matures in 20 years. He pays an annual premium of £4.95 per cent and receives an annual bonus of £26. How much will he collect when his policy matures?
2. A man takes out a whole life policy for £5,000. The annual premium is £4.50 per £100. What is the least number of years he would have to pay his premiums to cover the value of his policy?
3. A man insures his life for £3,500 and pays an annual premium of £105. What is the rate of insurance per cent?
4. If the premium rate on a life policy is $4\frac{3}{8}\%$ and a man pays £105 per year, for what sum is his life assured?
5. Find the annual cost to a man who has a twenty-five year policy for £10,000 life assurance if his premium is at the rate of £4.20 per hundred pounds, and on which he receives a tax rebate of 30p per £ of 3/5 of the premiums paid.

Exercise 84

Bankruptcy: A bankrupt is a person who has incurred debts which he cannot pay. His assets are less than his debts or liabilities. The amount which he can pay per pound of debt is his dividend and the people to whom the bankrupt owes money are his creditors.

1. A bankrupt owes £4,500 to Jones, £3,875 to Smith and £5,322 to Brown. From his assets he can only pay 37p in the £ dividend. What are his total assets?
2. A small firm is declared bankrupt with debts amounting £26,395. The owners can only pay a dividend of 68p in the £. What is the total loss the creditors suffer?
3. A bankrupt's debts amount to £6,675 and he can only pay a dividend of 18½p in the £. What are his assets?
4. A bankrupt's total assets are £6,384 and he can only pay a dividend of 32p in the £. What are his total debts?
5. A bankrupt has assets of £1,924 and liabilities of £5,200. What

dividend can he pay and how much will a creditor receive to whom he owes £4,073?
6. A shopkeeper goes bankrupt owing £6,500 and with assets of only £3,835. How much can he pay in the £?
7. A bankrupt has debts totalling £7,360 and he can only pay out £1,913.60. What dividend per £ does he pay and how much does a creditor lose to whom he owes £3,795?
8. A bankrupt paid a dividend of 35p in the £. One of his creditors received a total of £228.55. How much did this creditor lose?
9. The total assets of a bankrupt were £4,566.29. He incurred legal expenses of £563 which had to be paid in full and telephone and electricity bills of £79.41 and £127.88 which also had to be paid in full. If he owed his creditors a total of £14,600 what dividend can he pay them?
10. A bankrupt can only pay 40p in the £ to his creditors. If his total assets were £3,072 what were his liabilities?

Exercise 85

Foreign Exchange: In order that one country may trade with another it is necessary to evaluate the unit of currency in the first country in terms of the unit of currency in the second. Rates of exchange are published in the press and are fixed from day to day.

1. How many German marks will I receive for £43.50 if £1 = 7DM?
2. If I changed 1846 DM into British currency what will I receive if £1 = 7DM? (Answer to nearest £0.01).
3. The rate of exchange in Paris is 11.20 francs for £1. How much will I receive in francs for £143?
4. Find to the nearest £0.01 how much I will receive when I change 743 francs into British currency when the rate is 11.20 francs to £.
5. If I changed £736 in New York when the rate was £1 to 2.47 dollars how much would I receive in dollars?
6. After a holiday in America a tourist changed 203 dollars back into British currency. How much to the nearest £0.01 did he receive when the rate was £1 = 2.45 dollars?

7. In Amsterdam a holidaymaker changes a £25 travellers cheque into Dutch florins when the rate was £1 = 7.36 florins. How many did he receive?
8. A holidaymaker returning from Holland was given £1 for 7.82 Dutch florins. How much to the nearest £0.01 did he receive if he exchanged 93.50 florins?
9. How many pesetas will a traveller in Spain receive for £235 when the rate of exchange is 144 pesetas per £?
10. A returning holidaymaker from Madrid is given £1 for 150 pesetas. How much to the nearest £0.01 will he receive if he pays into the bank 6340 pesetas?

Exercise 86
Foreign Travel:
1. A British traveller to France changes £270 into francs, when the rate was £1 to 11.23 francs. He and his wife stayed for eleven days at a hotel where their daily expenses were 59 francs each. Other expenses per day for travelling, sightseeing and entertainment averaged 83 francs. Find how many francs he spent and how much to the nearest £0.01 he will return home with if the rate of exchange is still the same.
2. A Dutch family on an exchange holiday with a British family changed 2140 florins into £'s when the rate was £1 to 7.80 florins. During their 28 day stay they spent an average of £4.50 per day on food, and £3 per day for entertainment. For part of their stay they hired a car which cost them £58. Find the total they spent in £'s and how many florins they will return home with if the rate of exchange has altered to 7.26 florins for £1. (Answer to nearest florin).
3. A French tourist in Spain changes 1650 francs into pesetas when the rate was 1 franc for 12.9 pesetas. He intended spending 10 days but found his expenses were 2,500 pesetas per day. How many whole days can he stay and how many francs will he return home with if the rate is unchanged? (Answer to the nearest franc).
4. A British business man owes 67,200 francs to a firm in Paris. He

can pay direct or through an agent in New York. By making calculations in whole £'s find which method of payment is the cheaper and by how much. £1 = 11.23 francs, and £1 = 2.45 dollars and 1 dollar = 4.60 francs.

5. A British business man spends 3 days in Berlin where his expenses are 168 marks per day and then 2 days in Paris where his expenses are 264 francs per day. What do his expenses amount to? (Answer to the nearest £) if the rates of exchange are £1 for 6.90 marks and £1 for 11.40 francs.

CHAPTER 11. Rates, Income Tax, Gas and Electricity Bills.

EXERCISES 87 – 92

Exercise 87

Rates: This is the name given to the money charged by local authorities on all householders and property owners. The houses and property is given an assessed value and rates at so much in the £, are paid on this figure.

The money collected helps towards paying for such things as Education, Police, Parks, Libraries and other amenities in the district.

Calculate the answers to the nearest £0.01 where necessary:

1. Find the rates paid on a property with an assessed value of £230 in a district where the rate per £ is 84p.
2. A house is assessed at £135 and the rate per £ is £1.05. Find the rates paid.
3. In a certain district a rate of 19p in the £ is levied as the council houses rates subsidy. What is the total assessment of the district if the amount raised for this purpose is £57,950?
4. How much is paid in rates on a property with an assessed value of £185 if the rate per £ is 27½p?
5. If a rate of 75p in the pound raises a total of £13,500 what is the total assessed value of the district?
6. Find the rates paid on a property assessed at £46.50 in a district where the rate per £ is 69p.
7. In a small burgh assessed at £150,000 the rates collected total £24,375. What rate per £ has been levied?
8. What is the assessed or rateable value of a district in which the rate per £ is 50p and the amount collected in rates is £17,500?
9. A house has a rateable, or assessed value of £84 in a district where the rate per £ is 67½p. What is the total amount of rates paid on the house?
10. A district was assessed at £685,000 and the amount collected in

rates was £465,800. What rate per £ was levied?

Exercise 88

1. A householder has an assessed value on his house of £95 and he pays rates four times in the year. If the rate per £ is 93½p in his district how much does he pay in rates each quarter? (Answer to nearest £0.01).
2. The expenditure in a small town for a year was £210,500 and the assessed value of the area was £175,000. Find the rate per pound to the nearest £0.01 to meet the expenses. If, in fact, the rate was fixed at £1.24 in the £, what was the surplus?
3. In a certain town a rate of 90p in the £ raises a total of £57,150 in rates. What is the rateable or assessed value of the town?

   A householder in this town was assessed at £138 and appealed for a reduction. He was finally charged £113.40 in rates. Find the amount by which his assessment was reduced as a result of his appeal.
4. A district council charges a rate of £1.41 in the £ of rateable value. If a householder pays £14.10 four times a year in rates, what is the rateable value of his house?

   If this householder has two children at school and if 36p of the £1.41 per £1 received by the council goes towards education, how much is he contributing per week of the full year towards each childs education? (Answer to nearest £0.01).
5. In 1972 a house was rated at £160 and the rate per £ was fixed at £1.15. How much was paid in rates in 1972? In 1973 the assessed value was increased by 5% and the rate per £1 decreased by 10%. What is the new amount paid in rates?
6. The rateable value of a city is £6,598,000. Find the amount collected in rates if the rate per £ was fixed at 97p.

   The following year the rateable value was raised to £8,743,000 and the amount required in rates was £7,562,000. What rate per £ to nearest £0.01 was required?

7. In a certain district the rate per £ is 84p. Find how much a householder will pay in rates if his assessed value is £128.

   To raise the necessary money he invests a sum in a building society which gives him an 8% return. How much must he invest so that he will have enough interest left over to pay £12.30 for house insurance and £16.18 for feu-duty on his house?

8. The total assessed value of a small burgh was £834,000. The burgh built a sports centre costing £575,000 and this was paid for by ratepayers over a 5 year period. By how much in the £ would the rate per £ have to be increased during this period to cover the cost of the sports centre? (Answer to nearest £0.01).

9. The rateable value of a town is £978,000 and the rate per £ levied is £1.13. How much is collected in rates?

   If a total of £1,075,000 is required to meet expenditure how much is left over from the rates?

   A factory owner in the district becomes bankrupt and is unable to pay his rates. His property is assessed at £23,000. Will the rate surplus be able to cover this loss and how much will still be left?

10. A householder invests sufficient each year in £1 savings certificates to pay his rates and house insurances four years hence. If his rates one year amount to £146.88 when the rate per £ is 96p and his house insurances are £12.60 find the least number of savings certificates he must cash. A £1 certificate is worth £1.25 in 4 years. Find also the assessed value of his house.

Exercise 89

Income Tax: This is a tax on an individual's income. The amount paid depends on each person's circumstances and the amount allowed as tax free allowances:

The lower personal allowance for a single person is £595.

The higher personal allowance for a married man is £775.

Wife's earned income allowance is £595. (Husband and wife taxed separately).

Allowances for children not over 11 years of age £200; over 11 but under 16 £235; over 16 if receiving full-time education £265.

Life Insurance Relief: The amount of total premiums if less than £10: £10 if total premium is between £10 and £20 and one half of the allowable premiums if more than £20. Other sums which are added to the tax-free allowances are interest on mortgages, superannuation, professional fees. Any income left after deduction of allowances is taxed at the following rates. 30% on first £5,000 of taxable income. 40% on next £1,000 of taxable income. 45% on next £1,000 of taxable income. The rates go up to 75% of the remainder.

Example: A married man with 3 children aged 2, 4 and 14 earns £3,740 a year and has an income of £270 from investments. He pays £223 per year to a superannuation fund and other allowable expenses amount to £194. Find how much tax he should pay in a year.

Tax Free Allowances:

Higher personal allowance	£ 775
Childrens allowance	£ 635
Allowance for superannuation	£ 223
Miscellaneous	£ 194
Total tax free	£1827

Total income   £4010
Tax free income £1827
                £2183

Tax to be paid = 30% of £2183
               = £654.90

1. A director of a company has a total income of £9,600 per year from all sources. He is married with four children aged 8, 12, 17 and 19. The two older ones are still receiving full-time education. He pays £394 to a pension fund and £86 a year on life insurance premiums. His other permitted tax free allowances total £562. Calculate the total amount he should pay in tax.
2. A married man with two chil dren aged 3 and 6 has an annual income of £1746. He pays £48 per year in life insurance premiums.

In addition to his normal tax free allowances he has miscellaneous tax free allowances totalling £51. Find how much he should pay in tax each year.

If £13 is deducted each month from his salary for tax purposes, find by how much he has overpaid or underpaid at the end of the year.

3. A single man earning £156 per month has a life insurance policy on which the annual premiums are £82. In addition to his normal allowances he is permitted to deduct £73 as extra tax free allowances. Find the total tax he should pay in a year and his net monthly salary. (Answer to nearest pence).

4. A single man earns £2760 as a basic salary per year plus £958 commission for additional work. He also receives a dividend of £375 from stocks and shares investments each year. He receives his normal allowances and an additional £100 for a dependent relative plus miscellaneous allowances totalling £325. How much should he pay in tax each year. If he is married during the year and his personal allowance increases from £595 to £700 by how much will his tax be reduced?

5. A married man with three children aged 7, 13 and 17 and all at schooo has a salary of £4360 per year. His wife earns £973 for a part-time job. The man was given the normal allowances permitted, and in addition, £276 for superannuation £194 for mortgage interest and £485 for miscellaneous allowances. How much tax should be deducted from the man's salary and what is his net annual salary? How much tax does the wife pay and what is her net annual salary?

6. A single woman earns £1,563 per year. She receives additional tax free allowances of £329. How much tax should she pay in a year and what will be her net monthly salary?

7. A top salesman earns £5,000 per year plus commission of 2% on all sales over £10,000. He is married with three children aged 10 and twins aged 14. He pays life insurance premiums of £106 per year, superannuation of £308 per year, and interest on his house mortgage amounts to £237. Other allowable tax free expenses amount to £422. Find the tax he should pay in the year when his sales total £73,000 and find his net monthly salary.

8. The head of a tutorial department in a college receives a basic salary of £3,060. In addition he has a responsibility payment of £978 plus £117 for his teacher training qualification. Each year he has a dividend of £308 from investments, and his wife who is a part-time lecturer earns £1,200 per year. The couple have four children aged 12, 14, 16 and 19 all of whom are undergoing full-time education. Find the total income the man is allowed tax free if in addition to normal items he has superannuation payments of £258 per year, plus £174 interest on mortgage repayments and additional £369 miscellaneous items. Find also the couple's total tax payment for the year and net combined monthly salary.

9. A married man with one child aged 11 is the transport manager for a large firm, and earns £7,648 per year. He pays yearly insurance premiums on his life of £82 and his annual superannuation contribution amounts to £474. Other permitted tax free expenses amount to £223. Find the tax he should pay in a year.

   If, for tax purposes, £136 is deducted from his monthly salary find by how much he has underpaid or overpaid at the end of the tax year.

10. A depute town clerk is married and receives tax free allowances for two of his four children aged 12 and 15. His annual salary is £4,284 and he also has fees amounting to £620 for professional services. His wife is also employed and earns £1,780 per year. Find his total tax free allowances if they also include £279 interest on mortgage repayments; £302 for superannuation payments and £536 for miscellaneous items. What tax should he pay per year and what is his net monthly salary after tax deductions?

    How much tax will his wife pay per year if in addition to her normal allowance she has a total of £463 miscellaneous tax free allowances?

Exercise 90

Gas Bills: In houses where gas is used for cooking and heating, meters are installed which record in hundreds of cubic feet the volume of gas consumed. Since the gas supplied in different districts will vary in its calorific or heating value the volume of gas used is needed to calculate the number of **Therms** supplied to a house.

For **town gas**, that is gas produced in a gas works, 1000 ft^3 is approximately equivalent to 4-5 **Therms**.

Natural gas has a calorific value of almost double town gas.

The meter is read every 3 months (or quarter) and the bill for the quarter will depend on the initial and final readings.

Various methods of payment of the bills are used. For example
a) The first 35 therms at 19p per therm and the next 1000 therms at 12.2p per therm or b) A standing charge of £3.50 plus a charge of 9.4p per therm on all gas used.

Example: Successive readings on a gas meter are 4565 and 4673 (hundreds of ft^3) in an area where the calorific value of 1 ft^3 of gas is 450. Find number of therms used. Since 1 ft^3 produces 450 British Thermal unit and 1 Therm = 100,000 B.Th.U.

1,000 ft^3 produces 4.5 Therms.

Gas consumed = 467300 − 456500
              = 10800 ft^3
Therms used   = 10.8 x 4.5
              = 48.6 Therms

Find the number of Therms used from the following readings (all in hundreds of ft^3).

	Initial Reading	Final Reading	Calorific Value
1.	4165	4382	465 B.Th.U.
2.	3689	4071	940 B.Th.U.
3.	9864	9978	450 B.Th.U.
4.	5437	6015	490 B.Th.U.
5.	7689	8736	965 B.Th.U.

Exercise 91

Find the amounts to be paid in the following examples if the calorific value is 480 B.Th.U. in each case. The first **35 therms**, are charged at **21.3p per therm**; the next **1000** at **14.1p per therm** and remainder at **12.5p per therm**.

	Initial Reading (100 ft³)	Final Reading (100 ft³)
1.	6983	7146
2.	0369	1407
3.	5436	6079
4.	7218	8409
5.	5798	6847

Find the bills to be paid in the following examples if there is a **standing charge** of **£3.65** and a charge of **10.6p per therm**. The calorific value of the gas is 980 B.Th.U.

	Initial Reading	Final Reading
6.	7846	8439
7.	0758	1532
8.	1962	2706
9.	2439	3008
10.	1076	1904

Exercise 92

Electricity Bills: The amount of electricity used is determined by means of a meter with a dial counter or a digital display similar to gas meters. The unit on which the bills are calculated is the Kilowatt hour. A fairly high rate is charged for every habitable room in the house (for example 15 units per room per quarter) and the remaining units used at a lower rate.

Example: Find the bill for electricity for a 5 roomed house if the initial meter reading was 1436 and the final reading 2192. The charges are 15 units per room at 3.2p per unit and the remaining units at 0.92p per unit. Take answer to the nearest pence.

Units Used: 2192 − 1436 = 756

$$\begin{aligned} \text{Cost } 5 \times 15 &= 75 \text{ units at 3.2 pence} = £2.40 \\ &+ 681 \text{ units at 0.92p} = £6.27 \\ &\qquad\qquad\qquad \text{Total} \quad £8.67 \end{aligned}$$

Find the bills to be paid in the following examples if 15 units are charged per room at 3.2p per unit and all other units are charged at 0.92p. Answers to nearest pence.

	Initial Reading	Final Reading	Number of Rooms
1.	6049	7868	6
2.	5312	6751	4
3.	3725	5061	5
4.	8140	9465	3
5.	5864	7420	4
6.	0281	2024	4
7.	17834	18567	3
8.	9075	11362	5
9.	1684	3274	6
10.	8165	9937	4

CHAPTER 12. Logarithms. Multiplication. Division Powers. Roots and Mixed Examples using Logarithms.

EXERCISES 93 — 107

Exercise 93

Logarithms: These were invented by Baron Napier of Edinburgh in the 17th century. They are based on the laws of indices and are a great aid in the speedy calculation of multiplication and division examples.

The logarithm of a number consists of two parts.

a) The **characteristic**. This is an integer and depends on the position of the decimal point and is found by inspection.
b) The **mantissa**. This is a decimal part and is found from tables.

To find the logarithm of a number first express it in standard form.

Example: $5690 = 5.69 \times 10^3$

The index 3 is the **characteristic** and the **mantissa** .755 is obtained from row 56 and column 9 in the logarithmic tables.

Log. 5690 = 3.755

Logarithms of numbers greater than 1.

Find the logarithms of the following numbers.

1. 35.8
2. 863
3. 64.8
4. 9750.
5. 85000
6. 19.4
7. 475.
8. 85100
9. 635.
10. 3.14
11. 27.3
12. 4780.
13.. 60400
14. 8.37
15. 90.9.
16. 108.
17. 95.1
18. 466000
19. 78.1
20. 12.9

Exercise 94

To find the answer to a logarithmic question in normal number form we use the **antilogarithm tables**.

Look up the decimal part only and determine the position of the point from the characteristic.

2.874 is the logarithm of 7.48×10^2 = 748.

Use tables to find the numbers whose logarithms are:

1.	2.675	2.	1.638	3.	4.977	4.	3.841
5.	6.587	6.	5.938	7.	4.843	8.	3.636
9.	2.759	10.	0.568	11.	2.412	12.	5.713
13.	3.014	14.	4.815	15.	4.697	16.	3.019
17.	2.126	18.	3.716	19.	6.378	20.	5.648

Exercise 95

Multiplication: The first law of logarithms is multiplication. Look up the tables to find the logarithms of the numbers and **add**. The answer is then found in the antilogarithm tables.

Example: 4.23×783
= 3.31×10^3 = 3310

Number	Log
4.23	0.626
783	2.894
Expression	3.520

Find to 3 significant figures the results of the following products:

1. 63.4×265.
2. 718×21.9
3. 37.9×4.83
4. 51.8×758
5. 3.08×29.4
6. 4790×56.8
7. 6.1×31.7
8. 56300×4.51
9. $5.73 \times 318 \times 4.69$
10. $9.47 \times 21.6 \times 264$

Exercise 96

Division: The second law of logarithms is division. Look up the logarithms of the numbers and subtract. The answer is then found in the antilogarithm tables.

Example: 7250. ÷ 48.5
= 1.49 x 10² = **149**

Number	Log
7250	3.860
48.5	1.686
Expression	2.174

Find to 3 significant figures the results of the following divisions:
1. 954. ÷ 82.6
2. 49.6 ÷ 8.25
3. 524. ÷ 95.3
4. 7380. ÷ 94.7
5. 61800 ÷ 327.
6. 7240. ÷ 594
7. 1000 ÷ 39.5
8. 218 ÷ 2.06
9. 4720 ÷ 894
10. 16.3 ÷ 12.8

Exercise 97

Powers of Numbers: The third law of logarithms is finding the power of a number.

Look up the logarithm of the number in the tables and multiply by the power index.

Example: $(32.8)^2$
= 1.08 x 10³
= **1080**

Number	Logarithm	
$(32.8)^2$	1.516 x 2	3.032

Find to 3 significant figures the results of the following power examples.
1. $(53.3)^3$
2. $(6.98)^4$
3. $(891)^2$
4. $(3.77)^5$
5. $(29.2)^3$
6. $(45.7)^2$
7. $(7.36)^4$
8. $(10.7)^4$
9. $(8.36)^3$
10. $(7.08)^2$

Exercise 98

Roots of Numbers: The fourth law of logarithms is finding the root of a number.

Look up the logarithm of the number and divide by the root index.

Example:

$\sqrt[3]{21.5}$

$= 2.78$

Number	Logarithm	
$\sqrt[3]{21.5}$	$1.332 \div 3$	0.444

Find to 3 significant figures the results of the following roots:

1. $\sqrt{8.53}$
2. $\sqrt{64.5}$
3. $\sqrt[4]{709}$
4. $\sqrt[3]{52.7}$
5. $\sqrt{564}$
6. $\sqrt[3]{618}$
7. $\sqrt[4]{7580}$
8. $\sqrt[5]{312}$
9. $\sqrt[4]{51.9}$
10. $\sqrt[4]{100}$

Note: For the square root $\sqrt{}$; the root index is 2.

Exercise 99

Mixed Examples in Logarithms:

Find the value to 3 significant figures using tables of:—

$\dfrac{\sqrt{428} \times 7.59}{88.3} = 1.78$

Number		Logarithm
$\sqrt{428}$	$2.631 \div 2$	1.316
7.59		0.880
Numerator		2.196
Denominator		1.946
Expression		0.250

1. $\dfrac{63 \times 9.15}{86.4}$
                                 2. $\dfrac{85 \times 1190}{213. \times 5.42}$

3. $\dfrac{(28.5)^2}{9.84 \times 7.13}$ 4. $\dfrac{8160 \times 9610}{(453)^2}$

5. $\dfrac{(271)^3 \times 5.14}{(394)^2 \times 29}$ 6. $\dfrac{(36.4)^2 \times (5.28)^3}{(21.3)^3}$

7. $\dfrac{26.5 \times (21.7)^2}{(14.5)^2}$ 8. $\dfrac{42.8 \times \sqrt{93.4}}{(6.84)^2 \times 3.73}$

9. $\dfrac{29.1 \times \sqrt[3]{453}}{4.12 \times \sqrt{381}}$ 10. $\dfrac{218 \times \sqrt{829}}{57.1 \times (2.91)^2}$

Exercise 100
Logarithms of Numbers Less Than 1:
Example: $0.0942 = 9.42 \times 10^{-2}$

The -2 index is the characteristic and is written $\bar{2}$ (bar two) and it has a negative value.

The mantissa from the tables is .974 and it has a positive value.

Log. $0.0942 = \bar{2}.974$

Find from tables the logarithms of the following numbers.

1. 0.0453	2. 0.594	3. 0.0087
4. 0.0813	5. 0.000864	6. 0.004
7. 0.0365	8. 0.212	9. 0.00915
10. 0.561	11. 0.0472	12. 0.00829
13. 0.693	14. 0.0865	15. 0.947
16. 0.0135	17. 0.0767	18. 0.0959
19. 0.00314	20. 0.000229	

Exercise 101
Express with the decimal part positive.

Example 1: $\bar{2}.47 + \bar{1}.98 = -2 + .47 + (-1) + .98$
$= -3 + 1.45 = -2 + .45 = \bar{2}.45$

Example 2: $\bar{3}.86 - \bar{2}.39 = -3 + .86 - (-2) - (.39)$
$= -3 + 2 + .86 - .39$
$= -1 + .47 = \bar{1}.47$

Example 3: $\bar{2}.65 - \bar{1}.86 = -2 + .65 - (-1) - (.86)$
$= -2 + 1.65 - .86 = -2 + .79 = \bar{2}.79$

Example 4: $\bar{2}.75 \times 3 = (-2 + .75) \times 3$
$= (-6 + 2.25) = -4 + .25 = \bar{4}.25$

Example 5: $\bar{1}.962 \div 3 = (-3 + 2.962) \div 3 = \bar{1}.987$

Since $\bar{1}$ is not exactly divisible by 3 it is expressed as $-3 + 2$.
Express the following examples with the decimal part positive.

1. $\bar{2}.64 + \bar{1}.39$
2. $\bar{1}.84 + 0.36$
3. $\bar{4}.06 + 0.75$
4. $\bar{6}.92 - \bar{4}.87$
5. $\bar{3}.58 - 2.69$
6. $3.24 - \bar{1}.64$
7. $\bar{2}.17 - \bar{1}.88$
8. $0.72 - \bar{3}.64$
9. $\bar{2}.54 - 2.61$
10. $\bar{1}.26 \times 3$
11. $\bar{2}.046 \times 2$
12. $\bar{3}.867 \times 2$
13. $\bar{2}.605 \times 3$
14. $\bar{3}.965 \times 2$
15. $\bar{1}.645 \times 3$
16. $\bar{1}.279 \div 2$
17. $\bar{3}.925 \div 3$
18. $\bar{2}.394 \div 3$
19. $\bar{1}.665 \div 3$
20. $\bar{4}.762 \div 3$

Exercise 102
Use tables to find the numbers whose logarithms are:

1. $\bar{2}.925$
2. $\bar{4}.403$
3. $\bar{5}.241$
4. $\bar{1}.603$
5. $\bar{3}.079$
6. $\bar{4}.678$

7. $\bar{1}.767$ 8. $\bar{5}.236$
9. $\bar{2}.368$ 10. $\bar{1}.728$
11. $\bar{2}.084$ 12. $\bar{3}.246$
13. $\bar{2}.457$ 14. $\bar{3}.614$
15. $\bar{5}.317$ 16. $\bar{1}.402$
17. $\bar{6}.377$ 18. $\bar{4}.579$
19. $\bar{3}.263$ 20. $\bar{2}.092$

Exercise 103
Multiplication. Find Answers to 3 Significant Figures:
1. 0.0354 x 4.23   2. 0.0725 x 54.6
3. 0.836 x 6.74   4. 0.0954 x 0.835
5. 0.00085 x 0.93   6. 0.625 x 4.62
7. 8.36 x 0.0615   8. 0.00728 x 0.964
9. 9.25 x 0.0465   10. 23.8 x 0.724

Exercise 104
Division. Find Answers to 3 Significant Figures:
1. 0.653 ÷ 4.76   2. 0.729 ÷ 9.15
3. 0.958 ÷ 0.734   4. 7.82 ÷ 0.0537
5. 94.6 ÷ 0.0324   6. 83.6 ÷ 0.675
7. 4.85 ÷ 0.793   8. 0.574 ÷ 0.0963
9. 5.36 ÷ 794.   10. 12.8 ÷ 3140.

Exercise 105
Powers: Find answers to 3 significant figures.
1. $(0.273)^3$ 2. $(0.846)^2$
3. $(0.0483)^2$ 4. $(0.624)^4$
5. $(0.574)^3$ 6. $(0.0635)^4$
7. $(0.0072)^2$ 8. $(0.199)^5$
9. $(0.615)^3$ 10. $(0.584)^2$

Exercise 106
Roots. Find Answers to 3 Significant Figures:

1. $\sqrt{0.0792}$
2. $\sqrt[4]{0.0849}$
3. $\sqrt[3]{0.0694}$
4. $\sqrt{0.069}$
5. $\sqrt[3]{0.726}$
6. $\sqrt[4]{0.00976}$
7. $\sqrt{0.00245}$
8. $\sqrt[5]{0.0984}$
9. $\sqrt[3]{0.537}$
10. $\sqrt[4]{0.0403}$

Exercise 107
Mixed Examples

Example: Find the value to 3 significant figures of:

$$\frac{(0.698)^2 \times \sqrt{0.623}}{(0.924)^3 \times 4.19} = 0.116$$

Number	Logarithm	
$(0.698)^2$ $\sqrt{0.623}$	$\bar{1}.844 \times 2$ $\bar{1}.794 \div 2$	$\bar{1}.688$ $\bar{1}.897$
Numerator		$\bar{1}.585$
$(0.924)^3$ 4.19	$\bar{1}.966 \times 3$	$\bar{1}.898$ 0.622
Denominator		0.520
Expression		$\bar{1}.065$

Find the value to 3 significant figures of:

1. $0.496 \times 0.0384 \times 8.19$

2. $\dfrac{73.2 \times 0.0839}{48.7}$

3. $\dfrac{0.628 \times 5.82}{25.9}$

4. $\dfrac{28.4 \times 0.473}{0.569 \times 0.0835}$

5. $\dfrac{2.68 \times 347 \times 21.9}{213 \times 0.0938 \times 0.45}$

6. $\sqrt{\dfrac{7.46 \times 24.1}{6420}}$

7. $\dfrac{2.74 \times \sqrt{0.948}}{35.8 \times (2.74)^3}$

8. $\dfrac{500 \times \sqrt[3]{0.718}}{(22.1)^2}$

9. $\dfrac{\sqrt{0.0843}}{27.9 \times 6.42}$

10. $\sqrt{\dfrac{8.26 \times 0.439}{(3.18)^3}}$

CHAPTER 13. Area of rectangle. Area of triangle. Areas of various shapes.

EXERCISES 108 – 111

Exercise 108

Area of a rectangle. This is found by multiplying the length by the breadth.

$A = l \times b$. If length and breadth are in metres the area will be in square metres or m^2.

1. Find the area of a rectangular lawn which is 27 metres by 24 metres.
2. Find the area of a wall which is 72 metres long and 1.8 metres high. Find also the cost of painting it at 12p per m^2.
3. A rectangular card is 7.2 cm by 3.5 cm. Find its area and the total area of card required to make 1000 such cards.
4. A field is 240m long and 180m broad. What is its area in ares if $100m^2$ = 1 are?
5. A field is 327 m long and 165 m broad. What is its area in ares?
6. If a tile is 15 cm long and 10 cm broad what is the area of a passageway which requires 560 such tiles? (Answer in m^2).
7. A floor in a hall is 12 metres long and 8.6 metres broad. What is its area and the cost of staining it at 17p per m^2? (Answer to nearest pence).
8. The floor of a sitting room is 5.3 m long and 3.9 m broad. What is its area and what is the cost of a carpet to cover it at £4.20 per m^2 if an extra 10% has to be bought to allow for matching? (Answer to nearest £0.01).
9. A field is 130 metres long and 110 metres broad. Find its area and cost of fertiliser necessary to cover it at 31p per are.
10. The rectangular area in front of a shopping precinct is covered in paving stones 75 cm square. If 83 paving stones are required along its length and 29 along its breadth find the length and breadth of

the area in metres and the cost of the paving stones if they are 32p each.

Exercise 109
1. Find the breadth of a room if its area is 32.76 m² and its length is 6.3 m.
2. A lawn has an area of 345.6 m². Find its length if its breadth is 10.8 m.
3. A field contains 114 ares. Find its length if its breadth is 95 m.
4. A garden plot has an area of 89.61 m². If its length is 10.3 m, what is its breadth?
5. A hearth is made of tiles 10 cm square covering an area of 4800 cm². If its length is 3 times its breadth find the number of tiles used and length and breadth of hearth.

Exercise 110
Area of a triangle is equal to a half of the base times the altitude.

Find the areas of the following triangles.
1. Base 34 cm. Altitude 18 cm.
2. Base 2.9 cm. Altitude 2.1 cm.
3. Base 12.5 cm. Altitude 6.8 cm.
4. Base 9.6 cm. Altitude 11.9 cm.
5. Base 7.4 cm. Altitude 5.4 cm.
6. Base 11.2 cm. Altitude 12.9 cm.
7. Base 2.3 m. Altitude 1.6 m.
8. Base 1.9 m. Altitude 4.3 m.
9. Base 27.5 cm. Altitude 19.1 cm.
10. Base 3.6 m. Altitude 4.2 m.

Exercise 111

Find the areas of the following figures. (Not to scale) or the required lengths.

1.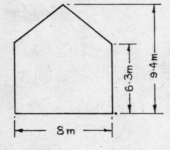

    Find the area

2.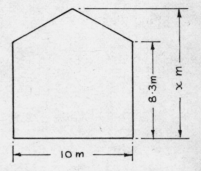

    Find x if the area of figure is 103.5 m^2.

3.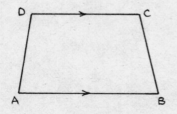

    If this figure represents the roof of a house, find the area of slates required if 10% extra is required for over-lapping.
    (Answer to 1 decimal place)

4. In this trapezium AB is parallel to DC
    AB = 9.3 cm   DC = 6.9 cm
    and the distance between AB and DC is 6.2 cm.
    Find area of figure.

5. 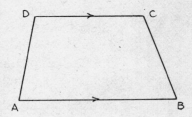 ABCD is a trapezium in which AB = 2.6 m, DC = 1.4 m and the distance between AB and CD is 1.8 m. Find area of figure.

6. 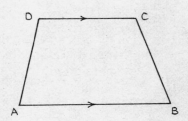 ABCD is a field in the shape of a trapezium. Find its area in ares if AB = 350 m, DC = 260 m and the distance between AB and CD = 240 m.

7. Find the area of a parallelogram, two of whose sides are 13¾ cm and the distance between them is 8½ cm.

8. A parallelogram has a pair of opposite sides of length 3 metres and the distance between them 1.8 m. Find its area.

9. 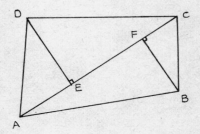 Find the area of quadrilateral ABCD if AC = 2.6 m and altitudes DE = 1.7 m and BF = 1.4 m.

10.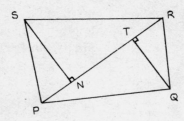

PQRS is a building site in which PR = 65.3 m, TQ = 29 m and SW = 30.2 m. Find the area of this site. (Answer to 1 decimal place).

11.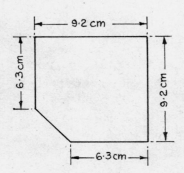

Find the area of this shape if the angles marked are right angles.

12.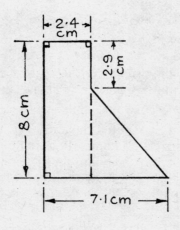

Find the area of this shape if the angles marked are 90°.

13.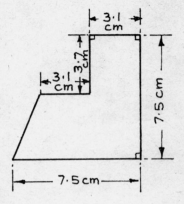

Find the area of this shape.

14.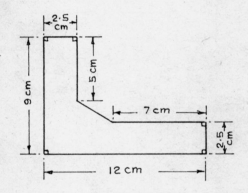

Find the area of this figure.

15.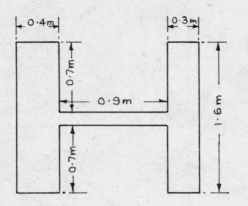

Find the area of this figure.

All measurements are in metres.

16.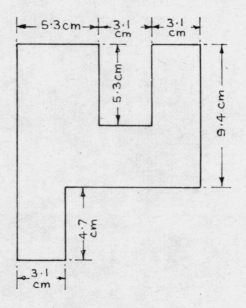

Find the area of this figure.

17.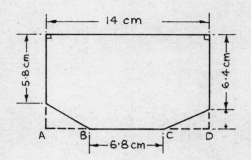

Find the area of this figure if AB = CD.

18.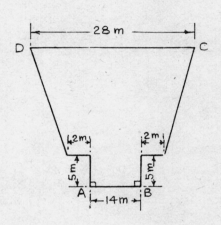

Find the area of this garden plot if AB is parallel to CD and the distance between them is 30 m.

19.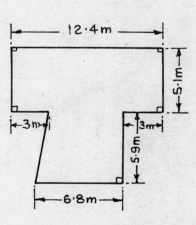

Find the area of this floor plan and the cost of staining it at 17p per m². (Answer to nearest pence).

20.

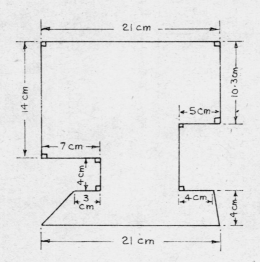

Find the area of this template.

CHAPTER 14. Circumference and area of circles and miscellaneous examples.

EXERCISES 112 – 118

Exercise 112

The **circumference of a circle.** This is given by the formula $C = 2\pi r$ or πd where r is the radius and d is the diameter of the given circle and π is the Greek letter pi denoting the ratio of circumference to diameter.

$\pi = \dfrac{22}{7}$ or 3.14.

Example: Find the circumference of a circle having a radius of 1.75 cm.

$c = 2\pi r = \dfrac{2 \times 22}{\cancel{7}} \times \cancel{1.75}^{.25} = 22 \times 0.5 = $ **11 cm**

Find the circumference of the following circles where $\pi = \dfrac{22}{7}$

1. Radius 3.5 cm.
2. Radius 5.25 cm.
3. Radius 28 cm.
4. Radius 9.8 cm.
5. Radius 9.45 m.
6. Diameter 5.25 cm.
7. Diameter 35 m.
8. Diameter 1.75 cm.
9. Diameter 63 m.
10. Diameter 15.4 cm.

Exercise 113

Find the circumference of the following circles to 2 decimal places where $\pi = 3.14$.

1. Radius is 3 cm.
2. Radius is 5 cm.
3. Radius is 10 cm.
4. Radius is 4.5 cm.
5. Radius is 6.5 cm.
6. Diameter is 16 m.
7. Diameter is 24 cm.
8. Diameter is 3.8 m.
9. Diameter is 6.7 m.
10. Diameter is 8.1 cm.

Exercise 114

1. A circular grass plot has a diameter of 42 m. Find the cost of fencing it at 17p per metre.
2. The radius of a wheel is 35 cm. How many revolutions will it make in travelling 3.3 km?
3. How many revolutions will the wheel of a car make in travelling 132 km if its diameter is 1.4 m? If the wheel is making 200 revolutions per minute find the speed of the car in Km/hr.
4. What is the distance round a running track if it has two straights of 90 m and two semi-circular ends of radius 35 metres?
5. A roller of radius 28 cm revolves 12 times when rolling the length of a lawn and 8 times when rolling the breadth. What is the length and breadth of the lawn?

Exercise 115

Find the radius or diameter given the circumference $R = \dfrac{C}{2\pi}$

Example: The circumference of a circle is 66 metres. Find its radius and diameter.

$$R = \frac{C}{2\pi} = \frac{66}{2 \times \frac{22}{7}} = \frac{\overset{3}{\cancel{66}} \times 7}{2 \times \cancel{22}} = \frac{21}{2} = 10.5 \text{ m}$$

$$D = 2 \times 10.5 = 21 \text{ m}$$

In examples 1 - 5 find the radius and in 6 - 10 find the diameter.
1. Circumference is 143 cm.
2. Circumference is 9.9 m.
3. Circumference is 220 cm.
4. Circumference is 13.2 m.
5. Circumference is 693 cm.
6. Circumference is 4.4 m.
7. Circumference is 33 cm.
8. Circumference is 17.6 m.
9. Circumference is 30.8 cm.
10. Circumference is 9.9 m.

Exercise 116

To find the area of a circle: $A = \pi r^2$. If the diameter is given find the radius first and then apply the formula.

Example: Find the area of a circle whose diameter is 28 cm.

Radius = 14 cm.

$$A = \pi r^2 = \frac{22}{\cancel{7}} \times \frac{\cancel{14}^2}{1} \times \frac{14}{1} = 22 \times 28 = 616 \text{ cm}^2$$

Find the areas of the following circles.
1. Radius 10.5 metres.
2. Radius 17.5 cms.
3. Radius 3.5 metres.
4. Radius 42 cm.
5. Radius 11.2 cm.
6. Diameter 2.8 metres.

contd. overleaf

7. Diameter $4\frac{2}{3}$ metres.
8. Diameter $5\frac{5}{6}$ metres.
9. Diameter 16.8 cm.
10. Diameter 52½ cm.

Exercise 117
To find the radius given the area

$$A = \pi r^2 \quad \text{hence} \quad r^2 = \frac{A}{\pi} \quad \text{and} \quad r = \sqrt{\frac{A}{\pi}}$$

Example: Find the radius of the circle whose area is 154 cm².

$$r = \sqrt{\frac{A}{\pi}} = \sqrt{\frac{154}{\frac{22}{7}}} = \sqrt{\frac{\overset{7}{154} \times 7}{22}} = \sqrt{49} = 7 \text{ cm}$$

Find the answers to one decimal place where necessary.

1. Find the radius of the circle whose area is 12.32 m².
2. Find the radius and circumference of the circle whose area is 2772 cm².
3. Find the radius of the circle whose area is 3773 cm².
4. Find the radius and circumference of the circle whose area is 1925 m².
5. Find the radius of the circle whose area is 246.4 cm².
6. If the area of a circle is $25\frac{2}{3}$ m² find its radius and diameter.
7. If the area of a circle is 30.8 m² find the diameter and the circumference.
8. Find the diameter of the circle whose area is 2860 cm².
9. Find the diameter and circumference of the circle whose area is 19.8 m².
10. Find the diameter of the circle whose area is 363 cm².

Exercise 118

1. A window of diameter 56 cm is cut from a square piece of glass 60 cm square. Find the area of 250 of these circular windows and the area of waste glass. (Answers in m^2).

2. Milk tops 3½ cm in diameter are cut from a roll 4 cm wide and 25 m long. How many milk tops can be cut from the roll and what is the area of the waste material?

3. In an entrance hall there is a circular design made of tiles. If the design is 3.5 m in diameter what is its area?

4. A children's circular paddling pond is 21 m in diameter. Find the area of tiles covering the bottom and the length of fencing round the edge.

5. A factory produces 10,000 circular plastic jam pot covers in a week. What area of material is required per week if the diameter of the covers is 7 cm and 15% extra material is required for wastage and overlaps? (Answer in m^2)

CHAPTER 15. Volume of cuboid, pyramid, cylinder, cone and sphere.

EXERCISES 119 − 125

Exercise 119

Volume of a Cuboid: This is found by multiplying the length times the breadth times the height.
V = l x b x h.
The surface area of a cuboid is the sum of the areas of the six faces.
Surface area = 2 (l x b + b x h + h x l).
Example: A block of stone measures 90 cm long by 60 cm broad and 50 cm high. Find its volume and total surface area.
V = 90 x 60 x 50 = 270000 cm^3.
Surface area = 2 (90 x 60 + 90 x 50 + 60 x 50)
= 2 (5400 + 4500 + 3000)
= 25800 cm^2.

1. A rectangular block is 2 m by 1.8 m by 1.3 m. Find its volume and total surface area.
2. A block is rectangular and measures 90 cm by 72 cm by 20 cm. Find its volume.
3. A sheet of hardboard measures 2.4 m by 1.6 m and is 5 mm thick. Find the total volume of 60 such sheets.
4. A household oil storage tank measures 1.6 m by 1.2 m by 1.2 m. Find its capacity in litres if 1 litre = 1000 cm^3. Find also its surface area.
5. A petrol storage tank measures 5 m long; 4 m broad and 2.6 m deep. How many litres of petrol will it hold if 1 litre = 1000 cm^3 and how many litres of paint would be required for its outside surface area if 1 litre covers 11 m^2 and it is only sold in litre tins?
6. A room measures 8 m by 7 m by 3 m. How many litres of air will this allow each of 36 pupils who occupy the room?

7. If a brick measures 20 cm by 10 cm by 7.5 cm how many would be required for a wall 45 m long, 2.5 m high and 30 cm wide?
8. A rectangular pond 72 m long, 36 m wide is frozen to a depth of 12 cm. How many m^3 of ice does the pond contain?
9. How many cubic metres of soil are removed from a trench 54 m long, 3 m deep and 1.4 m wide?
10. A cold water storage cistern measures 2.1 m by 1.2 m by 0.9 metres deep. If the water level is 10 cm from the top of the cistern find the area of the wetted surface and the number of litres contained.

Exercise 120
1. A room which contains 62400 litres of air is 5.2 m long and 2.4 m high. What is its breadth?
2. A swimming bath is 50 m long and 30 m wide. How many litres of water must be pumped in to raise the level by 5 cm?
3. A tank 9.5 m long, 4 m broad contains water to a depth of 1.65 m. If 5700 litres are added and the water now just reaches the top of the tank what is the tank's depth?
4. A tank which is 3.5 metres long and 3 metres wide contains 6720 litres. What is the depth of the water?
5. A reservoir which is roughly rectangular in shape is 250 metres long and 170 metres broad. On a certain day it contains 425000000 litres. What is the depth of the water and by how much will the level fall after 4 days supply have been drained off if 2500000 litres are used per day? (Answer to nearest cm).

Exercise 121
Volume of a Pyramid = 1/3 area of base x height.
1. Find the volume of a pyramid of height 12 cm and square base of side 7 cm.
2. Find the volume of a pyramid of height 15 cm and rectangular base 7 cm by 9 cm.

3. A pyramid has a triangular base of area 46 cm². If its height is 7.2 cm what is its volume?
4. A pyramid has a triangular base whose sides measure 5, 12 and 13 cm. If its height is 12.6 cm find its volume.
5. A pyramid has a volume of 150 cm³ and it has a square base of side 10 cm. What is its height?
6. Find the volume of a pyramid of height 6.2 cm on a square base of side 5 cm.
7. Find the volume of a pyramid on a square base whose height is 1.8 m and whose slant edge is 2.4 m.
8. A pyramid has a rectangular base measuring 1.2 m by 1.6 m and a slant edge of 2.6 m. Find its height and volume of the pyramid.
9. A right pyramid has a square base of side 4.8 cm and volume 38.4 cm³. Find its height.
10. A pyramid has a volume of 144 cm³ and a height of 24 cm. Find the area of its base and length of side of base to one decimal place if it is a square.

Exercise 122
The volume of a prism is the area of the uniform cross section times the distance between the ends.

1. A prism is of height 11 cm and its base is an equilateral triangle of side 6 cm. Find a) area of the triangular base and
   b) volume of the prism.
2. A triangular prism is of height 25 cm. Its base is a right angled triangle whose side are 5, 12 and 13 cm. Find the volume of the prism.
3. The depth of water in a swimming pool increases uniformly from 0.9 m to 3 m. The pool is 50 m long and 30 m wide. Find the volume of water contained in the pool.
4. A shallow trench is 30 m long and is shaped like an isosceles trapezium. The top is 67½ cm wide and the bottom is 37½ cm wide and the distance from top to bottom is 60 cm. Find the

volume of soil removed from the trench.

5. An assembly hall has a cross section shaped like a trapezium, the ends of which are 10 m and 7.5 m high and the length between them is 45 m. If the hall is 27 m wide what is the volume of air space contained in the hall?

Exercise 123
The **volume** of a cylinder is given by $V = \pi r^2 h$ where r is the radius of the base and h is the height of the cylinder.
The curved surface area is given by the formula $A = 2\pi rh$ and the total surface area of a cylinder is $A = 2\pi rh + 2\pi r^2$.
Example: Find the volume and curved surface area of a cylinder whose radius is 10.5 cm and height is 30 cm.

$$V = \pi r^2 h = \frac{22}{7} \times \frac{10.5^{1.5}}{1} \times \frac{10.5}{1} \times \frac{30}{1} = 10395 \text{ cm}^3$$

$$\text{Surface Area} = 2\pi rh = \frac{2 \times 22}{7} \times \frac{10.5^{1.5}}{1} \times \frac{30}{1} = 1980 \text{ cm}^2$$

1. Find the total surface area and the volume of a cylinder whose radius is 15 cm and whose height is 56 cm.
2. Find the volume and the curved surface of a cylinder whose radius is 4.2 cm and height is 6 cm.
3. Find the volume and total surface area of a cylinder of radius 35 cm and height is 90 cm.
4. A cylinder has a diameter of 10.5 cm and a height of 25 cm. Find its volume.
5. A cylindrical storage tank has a radius of 12 metres and a height of 21 metres. Find its capacity and number of litres of red lead required to paint its curved surface if 1 litre will cover 11m².
6. Water flows through a pipe of diameter 14 cm at a speed of 1 metre per second. How long will it take to fill a tank 9.8 m by 5.5 m by 4 m at this rate?
7. A pencil is 16 cm long and has a diameter of 0.7 cm. What is the

volume of 1000 such pencils?

8. The curved surface area of a cylinder is $29\frac{1}{3}$ cm². If its height is 7 cm find its radius.

9. The volume of a cylinder is 924 cm³. If its radius is 7 cm find its height.

10. A garden roller has a diameter of 56 cm. If its length is 1.5 m find the number of m² rolled when the roller makes 56 revolutions.

11. A "family sized" tin of soup has a diameter of 10.5 cm and a height of 12 cm. Find the volume of soup contained in the tin. Find also the area of paper required to cover the curved surface if 5% extra is required for an overlap.

    The factory produces 10,000 of these tins per day. What volume of soup in litres is produced in a 5-day week and what area of paper is required in m² to label all the tins?

12. The tank on a petrol tanker is approximately cylindrical in shape, 10 metres long and $2\frac{1}{3}$ m in diameter. Find its capacity in litres. (Answer to nearest litre).

13. A steel casting 30 cm in diameter and 56 cm long is packed in a box 60 cm long with square ends 35 cm wide. The unoccupied space is filled with sand. What is the volume of sand?

14. A tank is to contain 6,000 litres and it is cylindrical in shape. If its radius is 1.5 metres find its height to two decimal places.

15. A cylinder of diameter 5 cm contains water to a depth of 25 cm. A heavy metal object is dropped into the cylinder and is completely submerged. Find its volume if the water is now at a height of 32 cm.

Exercise 124

The volume of a cone is $\frac{1}{3}\pi r^2 h$ where r is the radius of the base and h is the height.

1. Find the volume of a cone of radius 6 cm and height 21 cm.
2. A cone has a diameter of 14 cm and a height of 30 cm. Find its volume.
3. Find the volume of a cone of radius 1.4 cm and height 13.2 cm.
4. A conical tent has a diameter of 5 m and a height of 3.5 m. If the tent is occupied by 5 men find the number of litres of air each man has to breathe.
5. Find the volume of a cone of height 3.0 cm and diameter 3.2 cm.
6. A cone has a volume of 6468 cm^3 and a height of 14.0 cm. Find the radius of its base.
7. A cone has a radius of 5 cm and a height of 275 cm^3. Find its height.
8. A cone has a volume of 297 cm^3 If its height is 3½ cm find the radius of its base.
9. Calculate the height of a cone of radius 7 cm and volume 2464 cm^3.
10. A pyramid has a square base of side 14 cm and a height of 30 cm. A cone has a diameter of 14 cm and height of 30 cm. Find the difference in their volumes.

Exercise 125

The volume of a sphere is given by $V = \frac{4}{3}\pi r^3$ and its surface area is $4\pi r^2$.

1. Find the surface area and volume of a sphere of radius 7 cm.
2. Find the surface area and volume of a sphere of radius 20 cm.
3. Find the capacity of a hemispherical bowl of radius 0.7 m in litres.
4. A solid sphere of diameter 14 cm is placed in a cylinder of height 14 cm and diameter 14 cm. The unoccupied space is filled with water. What volume of water is added?

5. A storage tank is cylindrical in shape with hemispherical ends. If its total length is 21 m and diameter is 3.5 m find its capacity in litres to the nearest whole number.
6. Find the outside surface area of the storage tank in question 5 and find the number of litres of paint required to give it two coats if 1 litre covers 11 m^2.
7. Find the volume and surface area of a sphere of diameter 12 cm.
8. Find the radius of a sphere of volume 1000 cm^3 taking the answer to one decimal place.
9. How many rubber balls of radius 3 cm can be made from 1 cubic metre of material? (Answer to 3 significant figures).
10. A sphere has a volume of 4950 cm^3. Find its radius to one decimal place and its surface area.

CHAPTER 16. Stocks and Shares.

EXERCISES 126 – 128

Exercises 126

Stocks and Shares: The necessary capital to start a business is often provided by interested members of the public. The capital is broken up into units known as shares and the individuals who provided the capital are the shareholders.

The nominal value of a share can be any value the company may decide – 50p, £1, £3.50 and they are only bought and sold in whole numbers.

The market value of the shares varies from day to day and this is the price at which they are bought or sold.

If the shares' market value is above its nominal value it is at a **premium**. If the market value is the same as the nominal value it is at **par** and if the market value is below the nominal value the share is at a **discount**. The shareholders normally receive a proportion of the company profits each year or a **dividend**.

When the **government** or a large corporation wishes to raise money they usually issue **stock** which has a **nominal value of £100**. This may be bought or sold in any quantity.

Stocks and shares are bought and sold for the public by **stockbrokers** who charge a small percentage for their services.

Example 1: Markworth 6% £2 shares stand at £3. Find a) the cost of 120 shares; b) the interest each year on the 120 shares; c) the number of shares that can be bought for £720 cash; d) the cash obtained by selling 350 shares.

a) Each share costs £3. 120 shares will cost £3 x 120 = **£360** cash.
b) The interest from each share is 6/100 of £2 = 12p
   The interest from 120 shares = 120 x 12 = **£14.40**.
c) Number of shares for £720 cash = 720/3 = **240**.
d) Cash obtained for 350 shares = 350 x £3 = **£1050** cash.

1. Find the cost of 320 £5 shares at £4.60 per share and find the annual income from them at 5½%.

2. Find the annual income from 650 £3 shares at 7% and find the total cost for them if their market value is £3.81.
3. Find the income from £1520 invested in £1 ordinary shares costing 81p each if the rate of interest is 6½%.
4. 200 £1 shares were bought when the price was 87p and sold when the price had risen to £1.26. What will be the increase in capital?
5. A sum of £900 was invested in 7% preference shares with a nominal value of £3 and a market value of £3.75. How many shares could be obtained and what income will be obtained each year?
6. Find the cost of 240 £2 shares at £2.36 and the income derived from them if the dividend is 12%.
7. 6% £2 preference shares produce an income of £108 for an investor who paid £2.50 per share. How much did he invest?
8. A shareholder had 160 £1 preference shares giving a dividend of 8%. He sold them at £1.35 and with the proceeds he bought 50p ordinary shares at 54p which were giving a dividend of 10%. Find his change in income.
9. A man sells 1650 50p shares standing at 80p and with the proceeds buys exactly 1800 shares in a second company. What is the market value of the second shares?
10. A man invested £1980 in £1 shares when they were standing at £1.37½. These shares gave a dividend of 9% and the investor later sold the shares at £1.60, investing them in 6% 50p shares at 45p. How many 50p shares did he buy and what was his change in income?

Exercise 127

Example: What sum must be invested in 6% stock at 103 to produce an income of £420?

£103 invested produces an income of £6.

$\dfrac{£103}{\cancel{6}} \times \dfrac{\cancel{420}^{70}}{1}$ invested produces an income of £420 = **£7210**.

1. Find the cost and income from £600 of 5% stock at 85.

2. Find the cost and income from £750 8% stock at 115.
3. Find the cost and income from £500 of 6% stock at 107.
4. Find the cost and income from £750 of 4% stock at 80.
5. Find the cost and income from £1825 of 6% stock at 104.
6. Find the annual income from investing £1743 in 6% stock at 83.
7. How much stock is obtained by investing £3432 in 7% stock at 104 and what annual income is obtained?
8. A man invested £4005 in a 5½% stock at 89. How much stock will he obtain and what will be his annual income?
9. How much 8% stock at 124 can be obtained by investing £8928 and what annual income will be obtained?
10. £10230 is invested in 4½% stock at 110. How much stock is obtained and what annual income is obtained?

Exercise 128
1. £12000 of a 5% stock was sold at 80 and the proceeds invested in a 7% stock at 64. Find the change in the investors annual income.
2. A man withdrew £3500 from a building society which gave interest of 6% and reinvested in stock at 140 and a dividend of 10%. On both incomes he had to pay income tax at the rate of 31p in the £1. Calculate the change in his net income.
3. A man received a legacy of £3425 and he invested part of it in 4% stock at 75. How much did he invest in this way if he had an income of £140 per year?

The rest of the legacy was invested in a building society which gave interest of 6½%. How much was invested in the building society and what income was obtained from this source?

4. Find the annual income from investing £300 in a 4% stock at 60.

How much should be invested in a 5% stock at 62½% to give the same income?

If a legacy of £7436 is to be invested in these two stocks so that the income from each is the same, find the amount of each investment.

5. A man held £900 of a 3¼% stock which he sold when the stock was at 95. If he invested the proceeds in a 5½% stock at 114, how much of the new stock would he obtain and what change would there be in his annual income?

6. A man holds £8000 of a 4½% stock which he sells when the stock is at 105. He invests three quarters of the proceeds in a 3½% stock at 84 and the rest in a 5% stock at 120.

   What sum did he obtain from the sale of his first stock and what is the change in his annual income?

7. £7560 is invested in 4% stock at 105 and £1752 invested in 5% stock at 73. If income tax is deducted from the interest at the rate of 15p per £1 find the net annual income.

8. A man sells £900 of a 5½% stock at 104 and invests the proceeds in a 4¼% stock at 78. What is the increase in his income?

9. £4200 is invested in 5¼% stock at 84. How much stock is obtained and what is the annual income?

   If the stock is later sold when it reaches 89½ find the gain in capital.

10. a) A speculator holds £8500 of a 4½% stock. What is his annual income?

    b) He sells this stock at 92 and invests half of the proceeds in a 6% stock at 115 and the other half in a 3½% stock at 68. What is the increase in his annual income?

CHAPTER 17. Estimation of Errors. Absolute Error. Relative Error Tolerance.

EXERCISES 129 — 136

Exercise 129

Errors in Measurement: Some measures are **exact** since they are found by counting; for example the number of apples in a box or the number of passengers on a bus. Others are **approximate** since they are found by measurement; for example the volume of liquid in a bottle or the time taken by a runner over 100 metres.

State whether the following quantities are exact or approximate:

1. Number of strokes in a round of golf.
2. The time taken to swim 100 metres.
3. The depth of water in a fish tank.
4. The number of points scored in a game of billiards.
5. The number of spectators at a football match.
6. The weight of soap powder in a packet.
7. The number of cigarettes left in a packet.
8. The number of apples in a crate.
9. The weight of the apples in a crate.
10. The number of litres of paint needed to paint the outside of a building.

Exercise 130

A given volume of liquid is stated to be 5.6 litres. This is its **true measurement**. This volume has been measured to the nearest 1/10 of a litre and 0.1 litre is the **least unit of measure**.

The **absolute error** permitted is one half of the least unit of measure and in this example is 0.05 litre and the **upper** and **lower limits** of the volume would be 5.65 and 5.55 litres respectively.

The **relative error** is the ratio of the absolute error permitted to the true measurement.

Relative error $= \dfrac{0.05}{5.6} = \dfrac{5}{560} = \dfrac{1}{112}$

The **percentage error** is the relative error expressed as a percentage.

The **tolerance** of a given measure is the difference between the greatest and least acceptable measures.

In the following examples state the least unit of measurement:

1. 53 seconds
2. 6.83 cm^2
3. 82 days
4. 349 km
5. 92 litres
6. 79.4 hectares
7. 82.9 quintals
8. 23.6 seconds
9. 45.63 tonnes
10. 6.942 km

Exercise 131

Find the absolute error in each of the following measurements.

1. 36 mm
2. 29.4 seconds
3. 47 kg
4. 96.8 litres
5. 1.96 gm
6. 30.2 ares
7. 123 days
8. 47.3 km
9. 7.342 cm
10. 85.37 kg

Exercise 132

Find the upper and lower limits of the true measurement in the following examples.

1. 65 seconds
2. 1.64 seconds
3. 68.4 cm
4. 0.375 mm
5. 9.8 litres
6. 24.93 cm^3
7. 2.7 kg
8. 0.95 g
9. 36.5 ares
10. 94.7 m^2

Exercise 133

Find i) the relative error; ii) the percentage error in each of the following examples to two significant figures.

1. 47 kg
2. 5.0 g
3. 0.9 g
4. 24 cm
5. 7.3 mm
6. 136 km
7. 12 seconds
8. 10.2 seconds
9. 6.5 litres
10. 80 cm^3

Exercise 134

Find the tolerance in the following examples where the upper and lower acceptable measures are as given.

1. 7.4 and 7.1 litres
2. 64.7 and 64.2 cm
3. 11.6 and 11.4 seconds
4. 9.26 and 9.24 g
5. 8.44 mm and 8.40 mm

Exercise 135

Find the upper and lower limits of acceptable measures where bilateral tolerance is given.

Example: (6.4 ± 0.1) litres

Upper limit accepted   =   6.5 litres

Lower limit accepted   =   6.3 litres

1. (9.2 ± 0.2) seconds
2. (56.8 ± 0.3) mm
3. (440 ± 2) m
4. (0.68 ± 0.01) seconds
5. (8.44 ± 0.03) g

Exercise 136

When measurements are added or subtracted the resulting absolute error is the **sum** of the absolute errors of all the measurements.

1. A runner was timed over two laps of a 400 m track and his times were noted as 56.2 seconds and 57.1 seconds. What would be his

fastest and slowest acceptable times for the 800 metres?
2. The frontages of three adjacent houses are 30.4 m, 26.7 m and 32.5 m measured to the nearest 1/10 metre. What is the greatest and least acceptable measurements of the total frontage?
3. A triangle has sides of 5.6 cm; 4.3 cm and 6.1 cm all measured to 0.1 cm. Between what limits will its perimeter lie?
4. An oil storage tank contains 450 litres and during successive weeks 56, 38, 87, 62 and 45 litres are run off, all measurements to the nearest litre. What is the least amount that can be expected to be left in the tank?
5. A swimmer was timed over four successive lengths of a swimming pool. These times were 15.6 seconds, 16.0 seconds, 15.7 seconds and 15.4 seconds. What would be his slowest and fastest times over four lengths if the times for each length had been measured to the nearest 1/10 second?
6. A motorist was timed over two stretches of road; the first being 100 m and the second 150 m. His respective times were 4.6 seconds and 6.2 seconds. What were (i) the greatest and smallest distances covered? (ii) the fastest and slowest times recorded for the total distance? (iii) the fastest and slowest speed in m/sec to two decimal places over the total distance.
7. A roll of foil for milk top contains a measured length of 2500 cm. On seven successive days 324 cm of the foil were used, all measures to the nearest cm. What is the maximum and minimum amount of foil that can be expected to be left?
8. In a holiday resort the hours of sunshine recorded to the nearest 1/10 hr for a certain week were 14.1, 12.6, 13.5, 10.9, 14.7, 8.2, 7.4. What was the greatest number of hours of sunshine recorded?
9. A wall is built with ten rows of bricks each brick with a measured width of 7.5 cm. The spaces between the bricks is cemented and measures 0.4 cm. What is the greatest and least height of the wall if the first row of bricks is on ground level?
10. A rectangular sheet of plywood is measured to the nearest cm and found to be 180 cm by 75 cm. What is its true area and what are its greatest and least expected areas? If two pieces 54 cm wide and

68 cm wide are cut off parallel to the 75 cm edge what is the smallest area of plywood that may be left?

CHAPTER 18. Probability and Tree Diagrams and Permutations

EXERCISES 137 – 139

Exercise 137

The **probability** of an event happening is the ratio
$$\frac{\text{number of favourable outcomes}}{\text{Total number of possible outcomes}}$$

Example: A letter is chosen at random from the word potato. What is the probability that it is (a) a vowel, (b) a t?

a) **Probability** $= \dfrac{3 \text{ vowels}}{6 \text{ letters}} = \dfrac{3}{6} = \dfrac{1}{2}$

b) **Probability** $= \dfrac{2}{6} = \dfrac{1}{3}$

1. Two dice are thrown. Make an array of the ordered pairs which will show on the dice. What is the probability;
   (a) that both numbers are the same?
   (b) the sum of the numbers is 6?
   (c) the sum of the numbers is at least 6?
   (d) what is the most likely total and what is its probability?

2. Three coins are thrown simultaneously. Make a list, (T for tail and H for head) of the possible outcomes. What is the probability of obtaining (a) two heads and a tail in any order? (b) two heads and a tail in the order T.H.T.?

3. From a complete pack of 52 playing cards, a single card is drawn at random. What is the probability that it is;
   (a) the ace of diamonds?
   (b) a six of any suit?
   (c) a black queen?
   (d) a black card?
   (e) a red ten or a black queen?

4. A bag contains 200 numbers for a prize draw, 10 of them being lucky numbers, and the numbers are not replaced. What is the probability of getting a lucky number if you are the first to try? After 75 numbers have been drawn (4 of them being lucky numbers) another try is made. What is the probability of obtaining a lucky number at the next try?

5. From a pack of 52 playing cards a double draw is made. What is the probability that the two cards are;
   (a) both kings?
   (b) both face cards (K. Q. J.)?
   (c) both spades?
   (d) both red?

6. From a pack of 52 playing cards two cards are drawn with replacement, i.e. the first card is replaced before the second is drawn. Calculate the probability that;
   (a) both cards are Jacks?
   (b) both cards are black?
   (c) the first card is a spade and the second is the ten of hearts?

7. A railway compartment holds 5 passengers on each side. What is the probability of a passenger;
   (a) sitting in a corner seat?
   (b) sitting in a corner seat and facing the engine?
   (c) sitting with back to engine and between two other passengers?

8. A cutlery box contains 20 knives and 15 forks. What is the probability of drawing;
   (a) a knife and fork in that order?
   (b) a knife and fork in any order?
   (c) two knives, followed by a fork?

9. Sixteen boys and sixteen girls enter for a mixed doubles tennis tournament. Two of the girls are twin sisters of Mr. A, one of the boys. What is the probability that Mr. A draws one of his sisters as a partner if he is the first boy drawn?

10. On a milk run, the number of litres of milk delivered per day to customers was:
    1 litre   —   33 customers

2 litres — 36 customers
3 litres — 22 customers
4 litres — 8 customers
5 litres — 3 customers
6 litres — 2 customers

What is the probability that a customer, picked at random, takes (a) 3 litres per day? (b) at least 4 litres per day?

Exercise 138

Tree Diagrams: Show in a tree diagram the number of different ways of arranging the letters a, b and c.

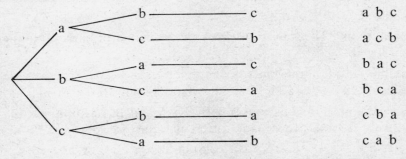

1. Show in a tree diagram the possible arrangements of boys (b) and girls (g) in a four children family. What is the probability that a family picked at random;
   (a) consists of two boys and two girls?
   (b) consists of g. b. b. g. in that order?
   (c) has a boy as the oldest member of the family?

2. X and Y are to play a snooker tournament and the first person to win two consecutive matches, or win a total of three matches, wins the tournament. Make a tree diagram to illustrate the number of ways in which the tournament can occur.

3. The following is the choice for a three course meal:
   First course: Soup, Fruit Juice or Tomato Juice.
   Second course: Fish or Roast Beef or Salad.
   Third course: Trifle or Ice Cream.

Make a tree diagram to show the number of different three course meals available.

4. Twenty-one boys enter a table tennis competition. How many boys would have to play off a preliminary round?
   Make a tree diagram to show the games played up to the final.
5. There are 4 spoons (S) and 3 forks (F) in a cutlery box and each item is drawn out without replacing.
   Make a tree diagram to show the possible ways of withdrawing the cutlery.
   What is the probability of obtaining the order S F S F S F S?

Exercise 139

Permutations: The ways in which a given number of unlike things may be arranged.

Example: The arrangements of the letters a, b and c are abc, acb, bac, bca, cab and cba i.e. there are six different arrangements.

1. There are four bus routes between A and B, but only three from B to A. How many different return journeys can be made?
2. If 8 matches in a football coupon have to be forecast correctly (win, draw or lose), how many forecasts would have to be made to have an all correct line?
3. If there are 8 horses in a race, how many ways are there of nominating the first 3 places?
4. If there are eight pupils on the school council, in how many ways can a chairman and a secretary be selected?
5. A combination lock consists of 4 rings with the digits 1, 2, 3, 4, 5 and 6. How many possible wrong combinations could this lock give?
6. A car license plate consists of two letters followed by three digits with the first digit not zero. How many different license plates can be made?
7. (a) How many 3 digit numbers can be formed from the six digits 2, 3, 5, 6, 7 and 8?
   (b) How many of these are less than 400?

(c) How many of these are multiples of 5 if repetitions are not allowed?

8. (a) Find the number of 4 letter words that can be formed from the letters of the word ENGLISH.
(b) How many of them contain only consonants?
(c) How many of them begin and end in a consonant?
(d) How many contain both vowels?
(e) How many begin with H and end with a vowel?

9. An ash, beech, birch, oak and elm tree have to be planted in a straight line. In how many ways can this be done?

10. Find the total number of positive integers that can be formed from the digits 2, 3, 6 and 7 if no digit is repeated in any one number.

11. The headmastership in 3 schools is vacant. If 8 men apply for the posts in how many ways can the appointments be made?

12. A toy consists of 5 different coloured rings which fit onto a rod. In how many ways can the rings be placed?

13. A father, mother and three children go to church. In how many ways can they take their places in the pew if the father sits at the head and his youngest child sits beside him?

14. Ten pupils of equal ability compete for two class prizes. In how many ways can the prizes be awarded? (Only one prize can be gained by a pupil).

15. Two brothers have 5 coats and 4 hats between them. In how many ways can they appear in a hat and coat each?

16. How many wrong ways are there of putting 5 letters in 5 addressed envelopes?

17. How many signals can be made by hoisting 5 different coloured flags when
(a) all are hoisted?
(b) any number are hoisted?

18. A small bookrack can hold 5 books of the same size. If 7 books are available in how many ways can the rack be filled?

19. How many even numbers of 3 digits can be formed from the

digits 2, 4, 5, 7, and 8 if repetitions are allowed?

20. Three passengers board a bus on which there are 5 vacant seats downstairs and 4 vacant seats upstairs.
In how many ways can they arrange themselves in the empty seats if;
(a) they all go downstairs?
(b) they sit anywhere?

CHAPTER 19. Statistics. Frequency. Distribution. Histogram. Mean Median Mode. Class Intervals. Class Boundaries. Cumulative Frequency Quartiles.

EXERCISES 140 − 149

Exercise 140

Statistics is concerned with the collection presentation and interpretation of facts.

This information may be about the weather or heights or weights of children of a certain age; preference in choice of television programmes attendances at football meetings and so on.

Different means are used to depict the given information; for example by means of a **pictograph**.

In this method the facts are represented by small drawings.

Another method is by means of a **line graph** as would be used for a hospital patient's temperature. The temperature would be taken at equal intervals of time and plotted on graph paper and the points joined with straight lines. The rise or fall of the line would be helpful to doctors and nurses.

A third type of graph is the **bar graph** where the information is represented by bars or rectangles of various heights or lengths.

The heights or lengths depend on the size and range of numbers involved.

Such graphs are commonly used to show imports and exports or study of output from factories.

The **pie chart** is another form of presenting information. A circle is divided into **sectors** like slices from a round cake or pie. The size of the angle at the centre of the sector depends on the fraction that each item is of the whole.

Example: A first year pupil has a 40 period week and the number of periods allocated to each subject is as shown. Draw a bar graph and a pie chart to illustrate this information.

French 6
Geography 3
P.E. 4
Religious Instruction
and Registration 2

English 8
History 3
Technical 5

Mathematics 7
Music 2

Bargraph

Pie Chart

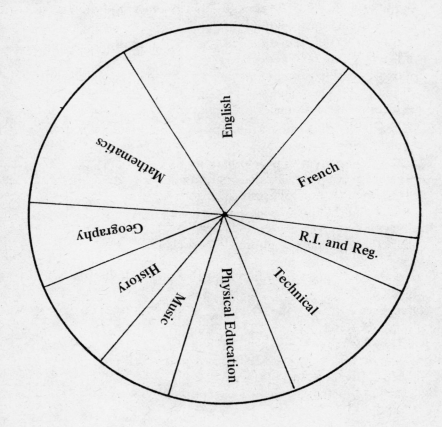

French allocation = 6 periods

Fraction of whole = $\dfrac{6}{40} = \dfrac{3}{20}$

Size of angle of sector = $\dfrac{3}{\cancel{20}_1} \times \dfrac{\cancel{360}^{18}}{1} = 54°$.

The same calculation is done for other subjects.

Draw a bar graph and a pie chart for each of the following examples:

1. A milk boy makes deliveries to 100 customers as follows:
   1 litre — 35 customers
   2 litres — 29 customers
   3 litres — 24 customers
   4 litres — 7 customers
   5 litres — 3 customers
   6 litres — 2 customers

2. A teacher's allocation of time for a week is made up as follows:
   Registration — 4 periods
   Teaching 1st Year Class — 6 periods
   Teaching 2nd Year Class — 3 periods
   Teaching 3rd Year Classes — 8 periods
   Teaching 4th Year Classes — 8 periods
   Teaching 5th Year Classes — 5 periods
   Supervision and Correction — 6 periods

3. The weekly budget for a family of five is as follows:
   Income — £35
   Expenses
   Rent and Rates — £5
   Milk — £2
   Meat and Fish — £5
   Cereals, Bread, Butter etc. — £8
   Coal and Electricity — £4
   Clothes — £2
   Entertainment, Cigarettes and Travelling — £6
   Saving — remainder

4. A school has a roll of 1,200 pupils and is made up in the following way.
   1st Year — 170 boys and 190 girls
   2nd Year — 150 boys and 150 girls
   3rd Year — 160 boys and 120 girls
   4th Year — 60 boys and 70 girls
   5th Year — 45 boys and 35 girls
   6th Year — 28 boys and 22 girls
   Divide each bar in your bar graph to make a compound bar graph

showing the composition of each year.
The pie chart should be drawn to indicate the numbers per year.

5. The numbers of 'O' grade presentations per pupil in the 4th, 5th or 6th year in a certain school were as follows.

1.	Presentation	13.	Pupils
2.	Presentations	16.	Pupils
3.	Presentations	14.	Pupils
4.	Presentations	14.	Pupils
5.	Presentations	19.	Pupils
6.	Presentations	32.	Pupils
7.	Presentations	52.	Pupils

Exercise 141

If the facts collected for a statistical example are obtained by **counting** them then the **variable** is said to be **discrete**.

For Example; the number of coins a group of men have in their pockets. If the facts are obtained by **measurement** then the **variable** is said to be **continuous**.

For Example; the month, by month weight of a baby over its first eighteen months.

State whether in the following examples the variable is continuous or discrete:

1. The number of tomatoes on a tomato plant.
2. The weight of tobacco used in making 1,000 cigarettes.
3. The tyre pressures in a selection of cars.
4. The number of candidates sitting 'O' level French in 1973.
5. The times taken by 100 runners in a cross-country race.
6. The number of hours of sunshine in a holiday resort in the month of July.
7. The times taken by employees to travel from home to their factory.
8. The number of passengers travelling on Corporation buses at a

given time each day for a week.
9. The barometric pressure readings of 100 places in and around the coast of Britain.
10. The number of eggs laid on a hen farm each day for a period of sixty days.

Exercise 142

If we count the number of persons in 100 passing cars and then tabulate the results, the number of times the number 4 (say) occurs, is called the frequency. The table constructed is called a **frequency table**.

No. in Car	Tally Marks	Frequency
1	ⅲⅰ ⅲⅰ ⅲⅰ 1111	19
2	ⅲⅰ ⅲⅰ ⅲⅰ ⅲⅰ ⅲⅰ 11	27
3	ⅲⅰ ⅲⅰ ⅲⅰ ⅲⅰ ⅲⅰ ⅲⅰ	30
4	ⅲⅰ ⅲⅰ 1111	14
5	ⅲⅰ 111	8
6	11	2

The **mode** is the number in a frequency table which occurs most often.

1. The marks scored by 100 children in a test were as follows.

7	6	7	6	5	4	5	5	4	5	3	2	7	3	6	6	5	7	6	8
8	9	10	7	4	7	9	6	7	7	2	10	6	5	4	4	3	9	7	5
6	4	4	4	5	9	6	6	3	5	6	5	4	6	6	2	5	3	6	6
5	7	5	1	7	6	7	5	7	2	4	6	8	7	2	6	7	7	9	3
5	8	8	6	5	6	4	6	8	9	6	8	6	8	3	4	4	4	5	7

Construct a frequency table and state which score is the mode.

2. A milk boy delivers the following numbers of bottles of milk to his customers.

2	1	1	4	3	2	5	1	3	3	3	1	3	6	2	5	3	2
2	4	5	2	3	2	1	1	1	3	2	2	4	4	2	1	2	2
2	2	1	1	3	2	6	1	3	3	1	1	2	1	1	2	2	
1	1	1	2	1	2	2	1	1	1	1	1	3	3	2	3	1	
2	2	2	3	2	3	1	2	3	1	3	1	3	3	2	4	3	
3	1	1	3	4	4	2	2	4	2	1	1	2	3	1	2	1	

Construct a frequency table and state what is the **modal number** of bottles delivered.

3. A survey of 100 families was taken and the number of children in each family was noted as shown.

2	4	4	2	3	2	1	3	3	4	1	5	3	1	1	2	4	5	1	1
0	2	3	3	5	3	3	0	6	2	3	3	4	2	2	4	6	4	3	4
5	6	4	1	2	4	3	1	3	7	5	3	3	4	2	6	3	0	1	7
2	1	4	4	3	2	1	3	2	2	4	1	2	3	4	3	6	3	5	2
1	2	3	3	2	2	2	5	5	6	1	3	5	5	3	2	4	4	2	5

Construct a frequency table and state the modal size of family from this survey.

4. Two dice were thrown 100 times and the sum of the numbers noted. Construct a frequency table and find the mode.

9	3	6	7	7	6	6	7	3	4	5	8	8	11	5	4	5	6	7	4
9	6	10	7	12	6	8	4	9	7	5	7	2	7	5	10	8	2	5	4
7	7	3	9	7	5	9	5	8	5	10	6	3	5	10	9	6	9	4	5
6	4	7	3	9	6	9	4	7	8	4	3	9	10	6	8	5	6	12	9
4	5	6	4	6	7	6	3	9	6	8	7	6	5	8	5	11	7	6	7

5. The following list gives the shoe sizes of a group of boys. Make a frequency distribution and find the mode.

9	6	6	7	6	5	6	5	6	5	4	5	4	5	6	5	7
4	8	6	8	2	7	8	7	4	6	6	6	7	9	3	7	4
3	7	9	5	6	6	4	5	2	5	7	9	4	7	2	5	10
7	10	4	4	5	7	8	6	3	10	4	3	6	6	4	6	
8	5	3	4	7	9	3	7	5	6	8	6	5	6	9	5	
4	2	6	3	6	6	7	9	7	7	6	4	7	5	6	3	
8	5	4	5	7	5	6	5	6	6	6	8	6	4	8	5	

Exercise 143

If the range of marks or observations is too large then they are grouped together into sets of equal size.

For example if the marks in an examination are percentages and range from say 6% to 98% they can be grouped 0-9, 10-19, 20-29 . . . and so on. A range of ten marks covers each group and this is the **class interval**. 0-9 is the **first class**.
10-19 is the **second class**.
20-29 is the **third class**.

In the example above the variable is discrete and the numbers at the bottom and top of each class are known as the **class limits**.

If the frequency table is made up from figures which have been rounded off then many of the measurements will actually lie between class limits of consecutive classes.

Example:
The heights of a group of 14 year old boys were taken rounded off to the nearest centimetre.

161	160	164	166	153	165	159	168
163	168	165	169	170	174	172	164
166	175	180	165	172	169	180	167
164	165	161	160	170	160	162	156
163	161	169	171	167	177	165	173

Make a frequency table of the heights with a class interval of 3 cm starting at 153 cm-155 cm.

The table will be as follows:

Height in cm	Tally Marks	Frequency
153 − 155	1	1
156 − 158	1	1
159 − 161	ℍℍ 11	7
162 − 164	ℍℍ 1	6
165 − 167	ℍℍ 1111	9
168 − 170	ℍℍ 11	7
171 − 173	1111	4
174 − 176	11	2
177 − 179	1	1
180 − 182	11	2

Since height is a continuous variable, some boys will have a **true height** which falls between consecutive class limits. A boy with a height of 164.4 cm will be rounded off to 164 cm and included in the 4th class.

In examples of this kind each interval is thought of as extending beyond its limits to the **class boundaries**. In example above the **class limits** of the **5th class** are 165-167 but the **class boundaries** are 164.5 and 167.5 cm. This class will include boys who are 164.5 cm and more but less than 167.5 cm.

The 5th class 165-167 cm is also the class containing the greatest number of boys. This is known as the **modal** class.

1. The following list is the marks of first year pupils in a Geography examination. Make a frequency table of 10 marks interval 1-10, 11-20, 21-30 etc. State which is the modal class

31	57	46	66	34	40	46	68	89	58	17	22
75	89	66	22	28	46	61	60	55	63	33	23
69	61	30	60	62	44	64	57	62	39	35	28
55	48	53	24	18	36	65	67	56	44	24	4
56	59	49	74	18	18	49	79	56	23	53	7
68	74	39	14	92	64	18	67	49	13	35	23
35	55	38	40	54	26	55	26	58	50	21	40
58	48	67	48	30	38	22	56	52	45	26	56
53	45	69	34	42	50	84	47	69	21	14	72
46	29	39	56	50	18	72	91	44	11	54	41

2. A sample of plants was taken from a batch that was planted at the same time and kept under the same conditions. Their lengths were measured to the nearest 1/10 cm and recorded as below. Put into groups:
20.5 − 21.5; 21.5 − 22.5 and so on.

25.5	24.6	25.2	23.9	23.1	24.8	24.2	23.6	24.2	27.3
24.5	23.5	24.2	22.9	24.7	24.0	26.0	24.4	23.2	22.9
22.6	24.1	24.0	25.3	25.3	24.3	24.3	24.7	23.6	25.1
20.6	23.8	22.8	23.1	23.9	24.8	24.0	26.3	23.7	22.1
23.5	24.5	23.5	24.6	23.3	24.2	23.4	23.6	23.9	22.7
25.6	24.3	22.1	23.5	24.0	25.9	23.9	23.4	23.7	25.1
23.8	25.8	24.9	25.0	25.4	24.9	23.6	24.1	22.0	23.7
23.0	25.2	23.8	21.2	23.6	25.4	21.8	24.9	26.4	25.0
22.7	24.0	26.8	24.7	24.2	22.5	24.2	24.6	24.1	22.6
21.6	24.3	22.4	23.0	24.7	24.1	27.2	22.8	23.9	23.7

Plants in the class 24.5-25.5 means they are 24.5 cm or more but less than 25.5 cm. From the frequency table state which is the modal class.

3. The following are the marks obtained by a group of pupils in a French test. Make a frequency table starting at 35 and using a class interval of 5 marks. Which is the modal class?

45	70	52	63	73	59	67	56	39	61
60	55	56	54	59	48	56	46	71	54
35	50	47	58	64	74	42	57	78	48
55	56	71	50	42	78	63	68	54	50
46	61	66	57	47	63	49	57	43	38
40	62	53	37	66	67	74	60	79	68
65	46	41	51	76	52	53	69	56	58
77	51	57	58	64	55	38	51	75	44

4. The times taken by 100 employees to travel from home to factory are as follows. All times have been rounded off to the nearest minute.

15	21	48	14	27	31	37	18	22	19
50	35	39	23	42	8	19	33	9	24
31	20	36	25	21	48	21	43	29	15
5	47	29	36	31	15	17	18	23	28
24	29	11	42	40	28	36	39	42	44
17	20	27	23	20	31	30	43	32	10
13	25	34	27	46	11	21	32	33	23
35	45	28	16	25	40	29	12	26	37
41	38	30	28	34	24	27	25	33	30
14	30	18	38	26	25	20	27	28	22

contd. overleaf

Arrange the times in intervals of 5 minutes 1-5, 6-10 and so on. State which is the modal class.

What are the class boundaries of the 6th class?

Exercise 144

Histogram: This is a diagram obtained from the frequency table and normally consists of vertical rectangles. The bases of the rectangles are made of equal widths and hence the heights of the rectangles will be proportional to the respective frequencies.

In the example above concerning the heights of 14 year old boys the bases of the rectangles may be marked in the following ways.

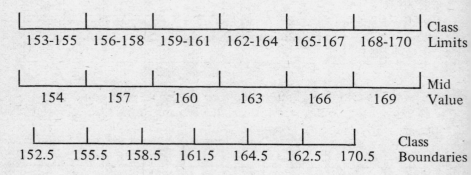

A FREQUENCY GRAPH

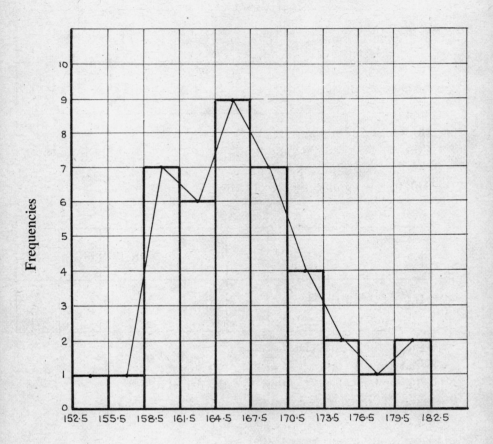

Heights in Centimetres

If the mid-pts of the tops of each rectangle are joined a curve or graph is obtained called the **frequency polygon**. The **frequency polygon** is also useful in calculations in frequencies and total number in a selection.

1. The times taken by a group of pupils to answer a mental arithmetic paper were as follows:

Time in Minutes	7	8	9	10	11	12	13	14
Number of Pupils	2	4	11	36	29	9	5	4

How many pupils were tested?
Construct a histogram and a frequency polygon to illustrate these results.

2. The following marks were obtained by pupils in a science examination:

Mark	11-20	21-30	31-40	41-50	51-60	61-70	71-80	81-90
Frequency	6	17	23	26	35	24	6	3

What is the modal class?
What are the class limits of the 3rd class?
What are the class boundaries of the 6th class?
Illustrate this information by constructing a histogram and a frequency polygon.
How many pupils were tested?
What is the probability that a pupil picked at random scored over 60?

3. The weights of 100 army recruits were recorded and are shown in the table below. (Weights were taken to the nearest kilogram):

Weight	60-61	62-63	64-65	66-67	68-69	70-71	72-73	74-75	76-77	78-79
Frequency	1	2	5	12	21	26	20	9	3	1

Illustrate this information on a histogram using class boundaries on the horizontal axis.

What percentage or recruits weigh 70 kg or more?

4. A poultry farmer kept a record of the eggs laid in a week by a number of hens.

The list is given below, and from it a frequency table should be made and then a histogram.

1	7	3	3	2	3	4	5	6	5	4	3	4	2	1
6	4	3	3	1	4	3	1	3	7	6	1	6	4	2
2	4	3	1	1	4	3	4	1	5	4	2	5	7	6
1	1	2	8	7	4	1	6	1	3	1	8	1	4	2
7	2	1	6	1	6	5	8	3	4	1	2	4	1	4
3	1	4	4	1	4	3	2	3	8	3	2	4	4	2
1	5	4	6	3	2	5	6	3	1	3	6	4	4	4
3	1	5	4	5	2	3	4	3	4	3	1	4	4	8

5. A passage was picked at random from a book and the number of letters in the first 200 words was counted.

This is recorded in the table overleaf.

No. of Letters	1	2	3	4	5	6	7	8	9	10
Frequency	4	25	60	48	33	12	4	7	5	2

What type of book would you expect this to be?
Construct a histogram and a frequency polygon.

Exercise 145

The **mean or arithmetic average** of a set of quantities is their sum divided by the number of quantities.

Example: Find the mean age of the following group of boys whose ages are: 12, 11, 12, 15, 10, 9, 13, 11, 12, 14, 12, 15.
Total = 146 years
Mean = $\frac{146}{12}$ = 12.16 years

1. In an examination the marks made by the ten candidates were 87, 74, 70, 67, 65, 62, 60, 59, 57 and 49. Find the mean mark.
2. During a twenty day period a traveller covered the following number of kilometres per day.
   97, 74, 96, 68, 82, 88, 54, 96, 27, 60, 33, 84, 35, 102, 71, 17, 55, 39, 78, 19.
   Find the mean distance covered per day to one decimal place.
3. Find the mean room temperature in a school where the readings for several rooms were as follows:
   16.3, 17.5, 16.4, 20.0, 16.9, 18.6, 16.5, 19.1, 19.4, 17.8, 16.7, 17.3, 19.4, 18.2.
   The readings being in degrees centigrade. Find answer to one decimal place.

4. The times taken by people using a public telephone were as follows:

   2 minutes 15 seconds   5 minutes 4 seconds
   3 minutes 28 seconds   6 minutes 11 seconds
   4 minutes 26 seconds   8 minutes 36 seconds
   4 minutes 5 seconds   7 minutes 45 seconds

   Find the mean time taken to the nearest second.

5. A tomato grower had 4 hot-houses producing tomatoes and the following is the weights of tomatoes produced per day by each hot-house over a 5 day period:

22 kg	28 kg	25 kg	36 kg	32 kg
26 kg	32 kg	39 kg	34 kg	26 kg
32 kg	35 kg	30 kg	29 kg	22 kg
33 kg	30 kg	28 kg	26 kg	36 kg

   Find the mean weight of tomatoes produced per day per hot-house to the nearest kg.

Exercise 146

If the number of quantities involved is very large then the sum of the products of quantities and their frequencies has to be taken. The symbol Σ (sigma) is used to denote a sum of given quantities.

Example: The following table shows the number of customers receiving various amounts of milk. Find the mean amount delivered per customer to the nearest 1/10 litre.

Number of Litres Delivered x	Number of Customers f	Frequency Times Customers f x
1	35	35
2	29	58
3	24	72
4	7	28
5	3	15
6	2	12

$\Sigma f = 100 \qquad \Sigma f x = 220$

Mean number of litres delivered $= \dfrac{\Sigma f x}{\Sigma f} = \dfrac{220}{100} = 2.2$ 1

The calculations can be greatly simplified by assuming a value for the mean and then making the necessary adjustment as follows:

Number of Litres Delivered x	Frequency f	x − A_m	f (x − A_m)
1	35	−2	−70 ⎫ −99
2	29	−1	−29 ⎭
A_m → 3	24	0	0
4	7	1	7
5	3	2	6 ⎫ +19
6	2	3	6 ⎭

Mean = Assumed mean + adjustment

$$= 3 + \frac{\Sigma f(x - A_m)}{\Sigma f}$$

$$= 3 + \frac{(-99) + (19)}{100} = 3 + \frac{(-80)}{100} = 3 - 0.8 = 2.2 \, l$$

1. A shopkeeper kept a record of the number of articles bought by customers during a day. Find the mean number of articles bought per customer.

Number of Articles x	1	2	3	4	5	6	7
Number of Customers f	5	12	36	43	34	17	3

What is the probability that the next customer will purchase exact-

-ly 5 articles?

2. The number of questions completed in an examination by a group of students were as follows.

Number of Questions Completed	3	4	5	6	7	8	9	10	11	12
Number of Students	2	3	6	18	35	51	41	24	15	5

Find the mean number of questions completed to the nearest whole number. If the students are expected to complete at least 7 questions to pass, what percentage of them will fail? (Answer to one decimal place).

3. A sample of packets of cereal (weight 400 grammes) was taken at regular intervals during a day. The packets were weighed to the nearest gramme and the results recorded as shown.

Weight in Grammes x	395	396	397	398	399	400	401	402	403	404	405
Number of Packets f	1	3	8	12	27	32	24	13	10	6	4

How many packets were tested?

What is the mean weight of the packets?

What percentage of the packets are under-weight? (Answer to nearest whole number).

4. A librarian recorded the number of books issued to members of the library over a three monthly period. The results are given in the table below.

Number of Books x	1	2	3	4	5	6	7	8	9	10	11	12
No. of Members f	16	20	64	80	124	160	142	108	75	51	44	16

Calculate the mean number of books taken out over this period to one decimal place.

Illustrate the figures by means of a histogram and a frequency polygon.

5. Two dice were thrown 200 times and the totals recorded. The results are given in the table below.

Total x	2	3	4	5	6	7	8	9	10	11	12
Frequency f	5	10	18	23	30	35	28	21	16	9	5

(i) Calculate the mean total thrown (Answer to one decimal place).

(ii) Make an array of all the possible pairs showing when two dice are thrown.

(iii) What is the theoretical probability of throwing a) a total of 4; b) a total of 7; c) a total of 11?

(iv) What would be the expected number of times that a) 4; b) 7; c) 11 totals would appear in 200 throws?

(v) How do the results from the table compare?

Exercise 147

If class intervals are used in the frequency table the mid values of each class are used in the calculation of the mean.

From the following table showing the distribution of marks scored by pupils in a French examination (maximum mark 50) calculate the mean.

Mark	Mid-Values x	Frequency f	f x
5-9	7	6	42
10-14	12	5	60
15-19	17	9	153
20-24	22	13	286
25-29	27	17	459
30-34	32	10	320
35-39	37	8	296
40-44	42	6	252
45-49	47	6	282

$\Sigma f = 80 \quad \Sigma fx = 2150$

$$\text{Mean} = \frac{\Sigma fx}{\Sigma f} = \frac{2150}{80} = 26.9$$

These calculations can also be simplified by assuming a value for the mean and making the necessary adjustments.

Mark	Mid-Values x	Frequency	x − A_m	f(x − A_m)	
5-9	7	6	− 20	− 120	
10-14	12	5	− 15	− 75	−350
15-19	17	9	− 10	− 90	
20-24	22	13	− 5	− 65	
25-29	27 A_m ←	17	0	0	
30-34	32	10	5	50	
35-39	37	8	10	80	340
40-44	42	6	15	90	
45-49	47	6	20	120	

$\Sigma f = 80 \qquad \Sigma f(x - A_m) = -10$

$$\begin{aligned}
\text{Mean} &= \text{assumed mean} + \text{adjustment} \\
&= 27 + \frac{\Sigma f(x - A_m)}{\Sigma f} \\
&= 27 + \frac{340 + (-350)}{80} \\
&= 27 - \frac{1}{8} = \mathbf{26.9}
\end{aligned}$$

1. The following table gives the frequency of the scores of candidates in a certificate examination in mathematics.

Score of Candidates	Frequency of Score
1-10	6
11-20	23
21-30	35
31-40	95
41-50	141
51-60	207
61-70	201
71-80	166
81-90	87
91-100	39

How many candidates were examined. Using the mid-values of each class, calculate the mean mark obtained by this group, taking your answer to one decimal place.

2. The following table gives the frequency distribution of weights of 200 women in a health club. (Weight to nearest kg).

Weights in kg	Frequency
40-44	4
45-49	6
50-54	9
55-59	23
60-64	36
65-69	48
70-74	30
75-79	22
80-84	9
85-89	8
90-94	5

Construct a histogram to illustrate these facts. Draw in the frequency polygon, and by using the mid-values of each class and an assumed mean, calculate the mean weight, to one decimal place.

3. An athletic club kept a record of the times recorded by its sprinters during a season. These are shown in the table below.

Time (Seconds)	10.2	10.3	10.4	10.5	10.6	10.7	10.8	10.9	11.0	11.1	11.2	11.3
Frequency	1	5	8	15	30	39	21	11	9	5	4	2

Make a histogram and a frequency polygon of these results.
How many recordings had been taken?
Calculate the mean time to 2 decimal places.

4. The following tables give the percentage absences of five boys' classes and five girls' classes over twenty weeks of a session. Make separate frequency distribution tables, histograms and frequency polygons. From the information calculate the mean percentage absence to one decimal place. (Use class intervals of 5% i.e. 0-4, 5-9, 10-14 etc.).
Which group in your opinion has the best attendance record over this period?

Boys' Classes

0	1	2	9	6	10	10	8	0	15	15	18	8	16	7	9	18	6	8	30
5	7	9	10	13	13	17	15	14	17	26	11	19	18	16	11	10	8	6	14
1	5	4	7	9	9	12	10	12	10	16	12	18	23	11	6	8	7	4	10
4	6	8	9	9	1	2	12	12	16	15	22	15	17	18	15	11	15	17	23
3	9	18	22	25	25	17	13	13	21	20	17	17	25	19	16	33	19	15	19

Girls' Classes

0	1	4	6	10	9	9	10	10	11	13	14	16	17	15	13	15	19	22	13
0	2	1	13	13	10	10	10	10	15	11	14	17	15	15	6	6	8	9	9
3	4	7	5	5	5	6	9	8	1	13	12	12	13	16	17	13	16	12	14
10	2	17	20	12	16	10	15	12	15	23	15	10	9	4	4	9	8	9	6
2	2	8	8	5	4	10	12	3	7	35	23	18	18	11	11	28	27	13	11

5. A garage owner is offering eight-fold pink stamps to customers who purchase a certain number of litres or more.

From the following table of customers supplied by the owner decide what the minimum number of litres is to qualify for the extra stamps.

Draw a histogram and calculate to the nearest whole number the mean number of litres sold.

Number of Litres	4	8	12	16	20	24	28	32	36	40
Customers Supplied	20	35	147	390	284	49	36	15	14	10

Exercise 148

Median: If a set of quantities is discrete then the **median** is the middle item, if the quantities are arranged in ascending order.

Example 1: Find the median height of the following group of boys measured to the nearest centimetre.
168, 143, 164, 147, 170, 167, 180, 169, 164, 178, 177.
In order these would be:
143, 147, 164, 164, 167, **168**, 169, 170, 177, 178, 180.
The 168 cm is the median since it has 5 quantities on either side.

Example 2: Find the median score in a class of 14 boys if the scores they made in an examination were:
18, 26, 39, 46, 48, 50, **52**, **53**, 55, 57, 59, 59, 68, 72.
The median score in this case lies between the 7th and 8th scores.
Median = $\dfrac{52 + 53}{2}$ = **52.5**

If the given quantities have been grouped in class intervals, it is only possible to indicate to which class the median belongs, if the number of quantities is large.

Upper and Lower Quartiles: If a set of quantities is arranged in ascending order the median divides the group into an upper and lower half. The **upper quartile** is the measure which divide the top half in two exact quarters and the **lower quartile** divides the bottom half in two exact quarters.

These measures may land exactly on an item or mid-way between two or if class intervals are used the class only may be known where the quartile lies.

Determine the Median and Upper and Lower Quartiles in the following examples:

1. The following marks were obtained by pupils in a French test.
   45, 43, 17, 82, 74, 67, 52, 60, 65, 72, 46, 86, 40, 38.
2. The following list gives the heights of 15 army recruits in centi-

metres.
190, 165, 174, 187, 166, 172, 184, 190, 166, 183, 180, 161, 175, 168, 182.

3. The following list gives the temperatures in different parts of Britain during a July day. (Measured in centigrade degrees).
16°, 24°, 20°, 18°, 17°, 16°, 24°, 25°, 17°, 18°, 21°, 19°, 16°, 20°, 19°, 25°, 21°, 22°.

4. The total number of hours of sunshine during the month of June from several holiday resorts is given below.
325, 363, 347, 359, 314, 329, 364, 372, 301, 317, 346, 358, 322, 336, 363.

5. The following list gives the weights of a group of 1st year boys in a school.
38, 54, 57, 39, 43, 45, 56, 39, 38, 47, 42, 44, 41, 49, 56, 51, 50, 39, 38, 54, 53, 51.
All weights in kg.

Exercise 149

If the number of scores or quantities is very large, it would be far too tedious to arrange them in order to determine the median. Instead, an extra column is added to the frequency distribution table — the **cumulative frequencies**.

These are obtained, by adding each frequency in turn to the total of those above it, as shown.

Number of Litres	Frequency	Cumulative Frequencies
4	20	20
8	35	20 + 35 = 55
12	147	55 + 147 = 202
16	390	202 + 390 = 592
20	284	592 + 284 = 876
24	49	876 + 49 = 925
28	36	925 + 36 = 961
32	15	961 + 15 = 976
36	14	976 + 14 = 990
40	10	990 + 10 = 1000

In the above example the **median** and **quartiles** can be found by inspection since these quantiles will be an exact number of litres.

The median which is denoted by Q_2 appears between the 500th and 501st frequency or the 500½. This will be one of the 390 customers who took 16 litres.

The lower quartile or Q_1 appears between 250 and 251 in the frequency list and this again will be included in the 16 litre group.

The upper quartile or Q_3 appears between 750 and 751 and will be one of the 284 customers who took 20 litres.

Two distributions with the same number of items can have the same mean but vastly different ranges.

Since between the upper and lower quantiles we have 50% of the observations, the **inter-quartile range** gives an **indication** or **measure** of the **spread** or **dispersion** of the marks.

To study this spread we calculate the **semi-interquartile range**. This is given by $\dfrac{Q_3 - Q_1}{2}$.

The figure obtained indicates the amount by which the middle 50% of the readings lie on either side of the median.

When class intervals are used in a frequency distribution **the upper class boundaries are plotted against the corresponding cumulative frequencies.** Where the variable is discrete, it may appear artificial using class boundaries instead of class limits, but it is more satisfactory to do so.

The following table gives the marks obtained in a test. Draw the cumulative frequency curve and from it determine the median, the upper and lower quartiles and the semi-interquartile range.

Mark	Frequency	Cumulative Frequency	Class Boundaries
1-10	2	2	0.5 – 10.5
11-20	10	12	10.5 – 20.5
21-30	18	30	20.5 – 30.5
31-40	16	46	30.5 – 40.5
41-50	21	67	40.5 – 50.5
51-60	24	91	50.5 – 60.5
61-70	18	109	60.5 – 70.5
71-80	6	115	70.5 – 80.5
81-90	3	118	80.5 – 90.5
91-100	2	120	90.5 – 100.5

The following points are plotted (10.5, 2); (20.5, 12); (30.5, 30); etc. Taking the cumulative frequencies of the first five classes we see that 67 scored 50 or less or 67 scored less than 50.5

OGIVE or CUMULATIVE FREQUENCY CURVE

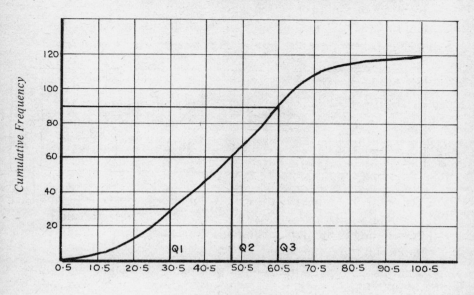

From the diagram the **median** Q_2 = 47.5.
The **lower quartile** Q_1 = 30.5.
The **upper quartile** Q_3 = 60.5.

The **semi-interquartile range** = $\dfrac{60.5 - 30.5}{2}$ = 15

1. The following table gives the shoe sizes of a group of school boys. By adding a cumulative frequency column, determine the median and upper and lower quartiles of the distribution.

   Calculate also the mean size of shoe to the nearest whole size.

Shoe Size	2	3	4	5	6	7	8	9	10
Frequency	7	15	31	36	42	33	16	14	16

2. The following table gives the marks obtained by pupils in an S.C.E. examination. Draw a cumulative frequency curve and from it determine the median upper and lower quartiles, semi-interquartile range and the mean, to one decimal place.

   What percentage of the pupils from this group will be given an A pass if 71 and over will qualify them for this grade? (Answer to one decimal place).

Mark	11-20	21-30	31-40	41-50	51-60	61-70	71-80	81-90	91-100
Frequency	5	10	25	50	53	60	84	62	31

3. Samples of a product (stated weight 60 g) were taken during a day and weighed to the nearest gram. The following results were obtained.

60	60	61	60	57	64	63	64	56	57	60	61	60	60	61	60	66	61	67	68
59	63	61	65	62	61	59	67	63	56	60	63	57	64	64	64	61	59	63	59
57	62	62	65	60	63	62	62	62	57	62	58	64	59	62	58	64	68	61	61
62	63	58	60	62	59	65	60	58	58	62	60	63	60	60	60	62	64	60	60
60	59	66	64	61	61	63	62	62	65	64	61	58	63	60	61	58	60	59	66

Make a frequency table and draw a cumulative frequency curve.
From it, determine the median and upper and lower quartiles.
What percentage of the sample is underweight?
If 10000 packets of the product are packed each day how many of them can we expect to be 60 gm in weight?

4. A selection of thirteen year old boys was taken and the amount by which they had increased in height over a year was recorded. This is given in the table below.

Increase in Millimetres	25-29	30-34	35-39	40-44	45-49	50-54	55-59	60-64	65-69
Frequency	5	13	34	55	73	62	45	10	3

(a) By taking mid-interval values, calculate the mean height increase, to one decimal place.
(b) Draw a cumulative frequency curve, and from it, determine the median, the upper and lower quartiles and the semi-interquartile range.

5. A random selection of householders was made in a city, and the amount of water they used per day was noted. The results are given in the table below.

Number of Litres Used	201-220	221-240	241-260	261-280	281-300	301-320	321-340	341-360	361-380	381-400	401-420
Frequency	7	35	99	147	181	360	251	167	66	31	6

(a) How many householders were investigated?
(b) Draw a cumulative frequency curve and determine the median and the quartiles and the semi-interquartile range to the nearest whole number.
(c) If the median is taken as the average consumption of water per householder and the city contains 150,000 households, how many litres are used per day?

ANSWERS TO EXERCISES

ANSWERS

Exercise 1
1. 19679
2. 53023
3. 134084
4. 94785
5. 45558
6. 86504
7. 97213
8. 226043
9. 60733
10. 57477

Exercise 2
1. 19144
2. 2127
3. 1581
4. 53478
5. 3909
6. 2127
7. 53478
8. 88
9. 2128
10. 29278

Exercise 3
1. 1257
2. 1718
3. 201
4. 2598
5. 1368
6. 3176
7. 681
8. 2331
9. 3602
10. 1412

Exercise 4
1. 15764
2. 184338
3. 418444
4. 8760
5. 159720
6. 2654292
7. 354888
8. 416787
9. 257760
10. 511119

Exercise 5
1. 79 R3
2. 142 R25
3. 235 R2
4. 139 R11
5. 85 R28
6. 119 R207
7. 91 R109
8. 245 R174
9. 72 R235
10. 162 R57

Exercise 6
1. 64.06
2. 131.473
3. 53.898
4. 26.442
5. 7.444
6. 60.633
7. 103.439
8. 123.453
9. 9.906
10. 19.787

Exercise 7
1. 7.28
2. 1.122
3. 0.7454
4. 8.558
5. 4.926
6. 0.252
7. 0.826
8. 2.559
9. 2.862
10. 0.556

Exercise 8
1. 10.38
2. 90.38
3. 11.86
4. 115.60
5. 1.425
6. 6.202
7. 9.476
8. 1.445
9. 1.770
10. 90.395

Exercise 9
1. 6.6464
2. 135.42
3. 194.682
4. 14.3936
5. 13.2130
6. 479.94
7. 9.065
8. 1.25664
9. 160.402
10. 4.3434

Exercise 10
1. 600
2. 3.18
3. 0.2007
4. 0.00924
5. 123
6. 0.3103
7. 0.07
8. 0.05023
9. 1737
10. 26.7

Exercise 11
1. £32.47
2. £49.96
3. £223.45
4. £93.23½
5. £168.46½
6. 30.72 kg
7. 241.86 kg
8. 32.809 kg
9. 32.561 kg
10. 2.094 kg
11. 54.00 l
12. 10.610 l
13. 19.533 l
14. 539. 03 m
15. 73.052 m

Exercise 12
1. £27.83½
2. £391.83½
3. £1041.18
4. £165.84½
5. £75.47
6. 38.541 kg
7. 0.3 kg
8. 47.151 kg
9. 8.015 kg
10. 5.133 kg
11. 10.46 l
12. 1.255 l
13. 113.87 m
14. 0.822 m
15. 35.042 m

Exercise 13
1. £741.06
2. £449.52
3. £21.06
4. £78.96
5. £251.28½
6. 102.76 kg
7. 291.87 kg
8. 674.10 kg
9. 762.72 kg
10. 3290.0 kg
11. 75.684 l
12. 144.60 l
13. 24. 192 m
14. 233.48 m
15. 4196.01 m

Exercise 14
1. £4.39
2. £12.26
3. £258.16
4. £3.64
5. £6.36
6. £252.06
7. 0.891 kg
8. 1.26 kg
9. 1.146 kg
10. 23.88 kg
11. 1.444 l
12. 1.416 l
13. 12.008 m
14. 3.5 m
15. 1.0046 m

Exercise 15
1. $2 \times 2 \times 5 \times 7$
2. $3 \times 3 \times 5 \times 7$
3. $2^2 \times 7 \times 11$
4. $2^2 \times 7 \times 13$
5. $2^3 \times 3^2 \times 7$
6. 5×29
7. $2 \times 3^3 \times 7$
8. 3×7^2
9. $3^2 \times 11 \times 13$
10. $2^3 \times 3^2 \times 11$

Exercise 16
1. 72
2. 48
3. 48
4. 90
5. 168
6. 360
7. 315
8. 168
9. 84
10. 385

Exercise 17
1. $\frac{2}{3}$
2. $\frac{2}{5}$
3. $\frac{3}{5}$
4. $\frac{1}{3}$
5. $\frac{1}{3}$
6. $\frac{8}{15}$
7. $\frac{2}{3}$
8. $\frac{2}{7}$
9. $\frac{2}{11}$
10. $\frac{2}{3}$

Exercise 18
1. $2\frac{2}{3}$
2. $1\frac{7}{8}$
3. $3\frac{5}{7}$
4. $6\frac{12}{13}$
5. $4\frac{1}{2}$
6. $2\frac{6}{11}$
7. $5\frac{1}{7}$
8. $2\frac{4}{5}$
9. $3\frac{1}{4}$
10. $4\frac{11}{15}$

Exercise 19
1. $\frac{15}{4}$
2. $\frac{21}{5}$
3. $\frac{56}{9}$
4. $\frac{38}{7}$
5. $\frac{23}{3}$
6. $\frac{64}{7}$
7. $\frac{21}{2}$
8. $\frac{93}{9}$
9. $\frac{32}{9}$
10. $\frac{49}{3}$

Exercise 20
1. $1\frac{1}{3}$
2. $1\frac{13}{24}$
3. $1\frac{3}{20}$
4. $1\frac{31}{60}$
5. $1\frac{23}{30}$
6. $3\frac{11}{12}$
7. $7\frac{13}{45}$
8. $16\frac{23}{24}$
9. $12\frac{17}{24}$
10. $11\frac{13}{18}$

Exercise 21
1. $\frac{1}{6}$
2. $\frac{11}{24}$
3. $4\frac{11}{24}$
4. $2\frac{1}{2}$
5. $\frac{5}{8}$
6. $2\frac{25}{42}$
7. $2\frac{11}{25}$
8. $2\frac{1}{24}$
9. $3\frac{11}{16}$
10. $1\frac{19}{21}$

Exercise 22
1. $3\frac{7}{12}$
2. $2\frac{1}{5}$
3. $2\frac{43}{48}$
4. $2\frac{23}{48}$
5. $3\frac{7}{12}$
6. $5\frac{15}{28}$
7. $4\frac{7}{24}$
8. $7\frac{7}{36}$
9. $5\frac{13}{24}$
10. $4\frac{1}{22}$

Exercise 23
1. 10
2. 8
3. 2
4. 40
5. $1\frac{1}{14}$
6. $26\frac{2}{3}$
7. $5\frac{1}{7}$
8. $\frac{4}{5}$
9. 12
10. 1½

Exercise 24
1. 2½
2. $9\frac{3}{5}$
3. $\frac{4}{9}$
4. 4
5. $5\frac{1}{3}$
6. 4
7. $\frac{21}{22}$
8. $\frac{68}{75}$
9. $1\frac{1}{7}$
10. $1\frac{1}{8}$

Exercise 25
1. $1\frac{1}{28}$
2. $\frac{7}{18}$
3. $\frac{4}{49}$
4. $8\frac{1}{9}$
5. $3\frac{8}{13}$
6. $2\frac{17}{28}$
7. $3\frac{13}{15}$
8. $\frac{99}{200}$
9. $1\frac{3}{10}$
10. 3¾
11. 1½
12. $4\frac{1}{12}$
13. 2
14. $2\frac{5}{12}$
15. $2\frac{3}{8}$

Exercise 26
1. £615.06
2. £4.08
3. 16½p
4. £3.25
5. 22p
6. 33p
7. 37½p
8. £4.48½
9. £34.41
10. 31½ litres
11. 80½ quintals
12. £6.23
13. 26¼ kg
14. £1.74
15. £11.97

Exercise 27
1. 9:28
2. 26:29
3. 19:14
4. 15:34
5. 3:5
6. 9:2
7. 129:200
8. 3:14
9. 31:50
10. 17:30

Exercise 28
1. 45p
2. 128 cm²
3. 35 kg
4. 2.1 l
5. £19.74
6. 48 l
7. 2.6 g
8. 40½ m²
9. 12.0 m
10. £88.0

Exercise 29
1. $\frac{7}{10}$
2. $\frac{11}{50}$
3. $\frac{9}{25}$
4. $\frac{9}{20}$
5. ¾
6. $\frac{21}{25}$
7. $\frac{23}{20}$
8. $\frac{2}{5}$
9. $\frac{7}{25}$
10. $\frac{19}{20}$
11. $\frac{5}{8}$
12. $\frac{1}{8}$
13. $\frac{1}{40}$
14. $\frac{3}{8}$
15. $\frac{1}{3}$
16. $\frac{2}{3}$
17. $\frac{7}{8}$
18. $\frac{1}{16}$
19. $\frac{3}{32}$
20. $\frac{9}{80}$

Exercise 30
1. 75%
2. 62½%
3. 31¼%
4. 220%
5. 66$\frac{2}{3}$%
6. 175%
7. 28%
8. 60%
9. 133$\frac{1}{3}$%
10. 68%
11. 50%
12. 25%
13. 62½%
14. 525%
15. 37½%
16. 125%
17. 0.75%
18. 250%
19. 12.5%
20. 8.75%

Exercise 31
1. £2.10
2. £1.12½
3. £25.85
4. £57.90
5. £12.12½
6. 19$\frac{1}{5}$ g
7. 0.352 l
8. 918.75 g
9. 1687.5 m
10. 44 minutes
11. 2408 metres
12. 307.5 g
13. 1.1 g
14. 0.510 l
15. 1.536 km

Exercise 32
1. 33$\frac{1}{3}$%
2. 19$\frac{4}{9}$%
3. 37½%
4. 60%
5. 12%
6. 87½%
7. 24%
8. 8%
9. 38%
10. 4%

Exercise 33
1. £1.72
2. £1120
3. 598
4. 420 litres
5. £6.00
6. 42 kg
7. 150 cm
8. £3200
9. 700 hectares
10. £150

Exercise 34
1. 5.79×10
2. 3.479×10^4
3. 6.24×10^2
4. 3.82×10
5. 9.7×10^5
6. 3.824×10^6
7. 2.16×10
8. 9.64
9. 5.68×10^4
10. 4.178×10^2
11. 5.39×10^{-1}
12. 6.84×10^{-2}
13. 2.95×10^{-3}
14. 1.02×10^{-2}
15. 9.246×10^{-5}
16. 1.78×10^{-1}
17. 6.482×10^{-2}
18. 1.98×10^{-2}
19. 6.285×10^{-2}
20. 4.53×10^{-5}

Exercise 35
1. 56.8
2. 0.07934
3. 7600
4. 3.951 and 4.0
5. 0.0058
6. 71.2
7. 418 and 420
8. 0.0405 and 0.040
9. 22
10. 30

Exercise 36
1. 2.984 and 2.98
2. 4.5743 and 4.57
3. 0.1852
4. 1.326
5. 1.20
6. 0.3453 and 0.34
7. 0.0067
8. 3.61
9. 73.084 and 73.1
10. 9.62

Exercise 37
1. 5
2. 7
3. 4
4. 9
5. 11
6. 100
7. 14
8. 30
9. 13
10. 25

Exercise 38
1. 24
2. 21
3. 35
4. 42
5. 54
6. 96
7. 93
8. 33
9. 38
10. 59

Exercise 39
1. $1\frac{1}{4}$
2. $\frac{7}{8}$
3. $\frac{4}{9}$
4. $2\frac{2}{5}$
5. $1\frac{5}{8}$
6. $2\frac{2}{3}$
7. $2\frac{1}{5}$
8. $3\frac{1}{3}$
9. $2\frac{3}{4}$
10. $1\frac{3}{5}$

Exercise 40
1. 1.41
2. 3.32
3. 5.39
4. 2.37
5. 4.15
6. 6.31
7. 9.18
8. 9.60
9. 6.37
10. 2.26

Exercise 41
1. 29.0
2. 82.0
3. 44.6
4. 237
5. 31.3
6. 838
7. 35.1
8. 28.5
9. 47.4
10. 647

Exercise 42
1. 0.982
2. 0.194
3. 0.284
4. 0.0122
5. 0.784
6. 0.221
7. 0.00929
8. 0.231
9. 0.351
10. 0.0868

Exercise 43
1. $2\frac{1}{3}$
2. 130
3. 4.11
4. 1.5
5. 4.22
6. 0.289
7. 7.69
8. 24.3
9. 0.025
10. 9.57

Exercise 44
1. 3.74
2. 2.64
3. 9.34
4. 20.3
5. 0.299
6. 0.111
7. 0.0446
8. 8.49
9. 1.53
10. 3.41

Exercise 45
1. 13.1 m
2. 84 m
3. 600 m and £103.50
4. 65 m and £464.75
5. 220 and 110 m
6. 60
7. 120 and 24 m
8. 31½ and 1½ m
9. 125 m and £20
10. 14.1 m
11. 5.25 cm
12. 8.4
13. 112½ secs.
14. 6
15. AD = 16 m and BC = 15 m Difference in guy ropes = 5 m

Exercise 46
1. 12, 15.
2. 1, -1.
3. 17, 13.
4. -5, -9.
5. -7, -11.
6. 8.6, 9.8.
7. 17, 21.
8. 22, 26.
9. 3a-6, 3a-8.
10. 57, 62.
11. 1/3, 1/9.
12. 27/4, 81/8.
13. 5/2, 5/4.
14. 243, -729.
15. -7/3, 7/9.
16. 1/2, 1.
17. -48, 96.
18. -27/2, 27/4.
19. 8/3, 16/9.
20. 9/8, 27/16.

Exercise 47
1. 7, 11, 15, 19.
2. 5, 8, 11, 14.
3. 7, 9, 11, 13.
4. 1/4, 1/7, 1/10, 1/13.
5. 13, 11, 9, 7.
6. 2, 6, 12, 20.
7. 2, 5, 10, 17.
8. 2/2, 4/3, 6/4, 8/5.
9. 4/1, 7/3, 10/5, 13/7.
10. 1½, 3, 5½, 9.
11. 5, 3/2, 7/9, 1/2.
12. 2, 6, 12, 20.
13. 1, 1/4, 1/9, 1/16.
14. 5, 7, 11, 19.
15. 1, 4, 27, 256.

Exercise 48
1. $n + 1$
2. $2n + 1$
3. $(n + 1)(2n + 1)$
4. $n + 4$
5. $11 - n$
6. $(n + 4)(11 - n)$
7. $n^2(n + 1)$
8. $5/3 \times 2^{n-1}$
9. $\dfrac{1}{n(n+1)}$
10. $9 - 2n$
11. $13 - 3n$
12. $5n - 3$
13. $3 \times (4/3)^{n-1}$
14. $50 \times (2/5)^{n-1}$
15. $n^2 + 1$

Exercise 49
1. 11
2. 13
3. 19
4. 54
5. 21
6. 29
7. 93
8. 55
9. 109
10. 221

Exercise 50
1. 10111
2. 101001
3. 100111
4. 111001
5. 110010
6. 1000011
7. 1001010
8. 1011101
9. 1110010
10. 1010101

Exercise 51
1. 1100
2. 10010
3. 11001
4. 10111
5. 100000
6. 11101
7. 11110
8. 100010
9. 100111
10. 101110

Exercise 52
1. 1010
2. 1000
3. 10010
4. 1001
5. 10001
6. 1010
7. 100010
8. 10101
9. 10000001
10. 1011101

Exercise 53
1. 101010
2. 1111
3. 1000001
4. 1000010
5. 101011111
6. 11100111
7. 110000001
8. 100011110
9. 10110100
10. 101001010

Exercise 54
1. 110 R1
2. 1011
3. 1001
4. 1000 R11
5. 110 R111
6. 11010 R101
7. 110 R11
8. 101 R100
9. 1101 R1000
10. 111 R11

Exercise 55
1. 48
2. 147
3. 73
4. 27
5. 56
6. 249
7. 198
8. 38
9. 578
10. 241
11. 615
12. 51
13. 86
14. 103
15. 181

Exercise 56
1. 1110 three
2. 321 four
3. 234 five
4. 121 eight
5. 220 three
6. 1221 four
7. 352 eight
8. 1240 five
9. 2021 three
10. 10032 four
11. 135 eight
12. 443 five
13. 244 five
14. 506 eight
15. 3034 five

Exercise 57
1. 8.05 m and 805 cm
2. 560 cm and 5600 mm
3. 8.056 Km and 8056 m
4. 850.96m and 85.096 Dm
5. 603.8 cm and 6038 mm
6. 4.625 Km
7. 34.96 m
8. 149.36 m and 0.14936 Km
9. 694.7 Dm and 6.947 Km
10. 632.7 m and 63270 cm

Exercise 58
1. 30027 cm^2
2. 90000 Dm^2
3. 10056 m^2
4. 76400 cm^2
5. 60493 Dm^2
6. 29.74 Dm^2
7. 5.7932 m^2
8. 7.659302 m^2
9. 158.76 ares, 1.5876 hectares
10. 5793.62 ares, 57.9362 hectares

Exercise 59
1. 69465 dm³
2. 0.80 m³
3. 6.974 cm³
4. 8.436 Dm³
5. 98.537 l
6. 900 l
7. 768.5 cl, 7.685 l
8. 53200 l
9. 168.43 l, 1.6843 Hl
10. 69.43 l

Exercise 60
1. 5630 Dg, 56300 g
2. 347.89 g, 3.4789 Hg
3. 9347 g
4. 53.69 Hg, 5369 g
5. 4700000 cg
6. 89376.2 g, 89.4 Kg
7. 9.647 Dg
8. 375.834 Kg, 37583400 cg
9. 5000000 mg
10. 12964.54 g, 12.96 Kg

Exercise 61
1. 9000 l, 18 days.
2. 44330 l
3. 66 times; 50 cm³
4. 350 customers
5. 3680 bags
6. 474.3 g
7. 1 hr. 40 min.
8. 1739 mm
9. 455.7 Dl
10. 4200 l

Exercise 62
1. £9.45
2. 253 km
3. £50.40
4. £15
5. 3.0 km
6. 5 hrs
7. 240 acres
8. 8 hrs $53\frac{1}{3}$ min.
9. 11 bricklayers
10. $21\frac{3}{5}$ years

Exercise 63
1. 60 days
2. 9 men
3. 9 days
4. $53\frac{1}{3}$ m
5. 60 days
6. 70 km/hr
7. 9 months
8. 9 men extra
9. 432 books
10. 7½ hrs.

Exercise 64
1. 99
2. 12.9
3. 7.2
4. 13 years
5. 54 runs
6. 2 min. 16 sec; 26.5 km/hr
7. 13.96 cm
8. £160.55
9. 185.3 and 11.7 cm
10. 1083

Exercise 65
1. 72 kg
2. 776
3. 61.6
4. £27
5. 51 years
6. £1397
7. 38p
8. 516 pts
9. 81 km
10. £47.27

Exercise 66
1. $89\frac{1}{3}$ km/hr
2. 65.9 km/hr
3. 24.4 km/hr
4. 72 km/hr
5. 57 km/hr
6. 43 km/hr
7. 61.7 km/hr
8. 57.6 km/hr
9. 48 km/hr
10. 64 km/hr

Exercise 67
1. £16.00
2. £2970
3. £19.36
4. Loss of £8
5. £133.20 Gain
6. £3 Loss
7. £16
8. £33.20 Loss
9. £37.20 Gain
10. £218.50 Gain

Exercise 68
1. 25% Gain
2. 20% Gain
3. 37½% Loss
4. 25% Profit
5. 20% Profit
6. $14\frac{2}{7}$% Profit
7. 25% Loss
8. 17½% Loss
9. $9\frac{1}{11}$% Loss
10. $14\frac{2}{7}$% Loss

Exercise 69
1. $20\frac{5}{6}$%
2. $16\frac{2}{3}$%
3. $22\frac{2}{3}$%
4. 324%
5. £200.30 and 33.4%
6. $22\frac{1}{12}$% Loss
7. $33\frac{1}{3}$% Gain
8. 25% Gain
9. $37\frac{1}{7}$%
10. 47.2%

Exercise 70
1. £10.92
2. £25.30
3. 56p
4. £7.70
5. £20.00
6. £1050
7. £2975
8. £2.07
9. £81.90
10. £1.36½

Exercise 71
1. £432
2. £16
3. £1.26
4. £9.00
5. £10
6. £1.60
7. £6.00
8. £1728
9. £27.00
10. £2.00

Exercise 72
1. £6750
2. 115%
3. £70
4. £4.08
5. £656.25
6. £20.70
7. Loss of 5%
8. £183.75
9. £6.60
10. 75% Gain

Exercise 73
1. £11.70
2. £17.60
3. 96 dollars
4. 7 francs
5. £36
6. £1.35
7. 210 dollars
8. £21
9. $28\frac{1}{8}$ francs
10. 72 francs

Exercise 74
1. £500.80
2. £579.60
3. 226.95 dollars
4. 826.80 francs
5. £5841

Exercise 75
1. £6.80
2. £28
3. 78.75 dollars
4. 27 francs
5. £9.58
6. 39.20 dollars
7. £14.06
8. £23.28
9. £24.32
10. £18.60

Exercise 76
1. £50.56
2. £73.23
3. £30.75
4. £36.45
5. £187.28
6. £27.05
7. £43.93
8. £128.83
9. £130.05
10. £119.25

Exercise 77
1. £731.16
2. £297.65
3. £2129.41
4. £616.26
5. £677.91
6. £967.99
7. £609.43
8. £614.85
9. £446.68
10. £735.78

Exercise 78
1. 33275
2. £794
3. £304
4. 774282
5. £483
6. £16748
7. 94814 toys
8. £11313
9. 19520 tonnes
10. 185 kg

Exercise 79
1. £121.60
2. A: £56.87½
   £34.56
   B: £70.58
3. £78.85
4. a) £4.86
   b) £14.58
   c) £42.07½
5. a) £3.19
   b) £5.44
   c) £1.39
6. £1288
7. $28\frac{1}{3}\%$
8. 25%
9. 9.3%
10. £226.32 and £215.01

Exercise 80
1. £129.54
2. £129.21
3. £9.57½
4. A: £7.24
   B: £5.68
   C: £10.18
   D: £9.62
   Manager: £20.47
5. £139.32
6. £82.85
7. £1912
8. £1390
9. £108.57
10. £43.80

Exercise 81
1. £43.75
2. £6.75
3. £78.50
4. £191.20
5. £11.41
6. £22.00
7. £84.45
8. £380
9. £5.85
10. £136.57½

Exercise 82
1. £2.85
2. £30.49
   Difference £10.91
3. £47.70
4. £31.50
5. £1575
6. £61762.50
7. Total Premium £1242
   Share 1/9
8. £12800
9. £137.50
10. £17.10

Exercise 83
1. £2500
2. 23 years
3. 3%
4. £2400
5. Rebate £75.60
   Annual Cost £344.40

Exercise 84
1. £5067.89
2. £8446.40
3. £1234.87½
4. £19950
5. 37p per £; £1507.01
6. 59p per £
7. 26p per £; Loss of £2808.30
8. £424.45
9. 26p per £
10. £7680

Exercise 85
1. 304.50 marks
2. £263.71
3. 1601.60 francs
4. £602.05
5. 1817.92 dollars
6. £82.86
7. 184 florins
8. £11.96
9. 33840 pesetas
10. £42.27

Exercise 86
1. 2211 francs and £73.11
2. £268 and 46 florins
3. 8 days; 100 francs
4. £5984 direct £5963 by New York. Save £21 by paying through agent.
5. £73 in Berlin. £46 in Paris. Total £119.

Exercise 87
1. £193.20
2. £141.75
3. £305000
4. £50.87½
5. £18000
6. £32.08½
7. 16.25p per £
8. £35000
9. £56.70
10. 68p per £

Exercise 88
1. £22.21
2. £1.21 Surplus £6500
3. £63500 Reduction of £12
4. £40; 14p per week
5. £184; £173.88
6. £6400060; £87
7. £107.52; £1700 is invested
8. 14p per £1
9. £1105140 Surplus £30140; Yes Amount left £4150
10. Assessed Value £153; Cashes £128; £1 Saving Certificate

Exercise 89
1. £2287.45
2. £148.80 Overpaid by £7.20
3. £348.90; £126.92
4. £921.90; £31.50
5. £579; £3781 £113.40; £859.60;
6. £191.70; £114.28
7. £1138.50; £426.79
8. £2576; £747.60 Net combined salary per month £409.61
9. £1860 Underpaid by £228
10. £2362 tax free £762.60 tax free £345.11 net monthly salary. Wife pays £216.60 in tax.

Exercise 90
1. 101 Therms
2. 355 Therms
3. 51.3 Therms
4. 283 Therms
5. 1010 Therms

Exercise 91
1. £13.59
2. £73.02
3. £46.03½
4. £83.13
5. £73.51½
6. £65.25
7. £84.05
8. £80.93
9. £62.76
10. £89.66

Exercise 92
1. £18.79
2. £14.61
3. £14.00
4. £13.22
5. £15.49
6. £17.41
7. £7.77
8. £22.75
9. £16.68
10. £17.67

Exercise 93
1. 1.554
2. 2.936
3. 1.812
4. 3.989
5. 4.929
6. 1.288
7. 2.677
8. 4.930
9. 2.803
10. 0.497
11. 1.436
12. 3.679
13. 4.781
14. 0.923
15. 1.959
16. 2.033
17. 1.978
18. 5.668
19. 1.893
20. 1.111

Exercise 94
1. 473
2. 43.5
3. 94800
4. 6930
5. 3860000
6. 867000
7. 69700
8. 4330
9. 574
10. 3.70
11. 258
12. 516000
13. 1030
14. 65300
15. 49800
16. 1040
17. 134
18. 5200
19. 2390000
20. 445000

Exercise 95
1. 16800
2. 15700
3. 183
4. 39300
5. 927.
6. 272000.
7. 193.
8. 254000.
9. 8530.
10. 54000.

Exercise 96
1. 11.6
2. 6.01
3. 5.50
4. 78.0
5. 189.
6. 12.2
7. 25.3
8. 106.
9. 5.28
10. 1.27

Exercise 97
1. 152000
2. 2380
3. 794000
4. 759
5. 24800
6. 2090
7. 2940
8. 13100
9. 583
10. 50.1

Exercise 98
1. 2.92
2. 8.03
3. 5.16
4. 3.75
5. 23.8
6. 8.51
7. 9.33
8. 3.16
9. 2.69
10. 3.16

Exercise 99
1. 6.65
2. 87.7
3. 11.7
4. 383.
5. 22.8
6. 20.3
7. 59.3
8. 2.37
9. 2.78
10. 13.0

Exercise 100
1. $\bar{2}.656$
2. $\bar{1}.774$
3. $\bar{3}.940$
4. $\bar{2}.910$
5. $\bar{4}.937$
6. $\bar{3}.602$
7. $\bar{2}.562$
8. $\bar{1}.326$
9. $\bar{3}.961$
10. $\bar{1}.749$
11. $\bar{2}.674$
12. $\bar{3}.919$
13. $\bar{1}.841$
14. $\bar{2}.937$
15. $\bar{1}.976$
16. $\bar{2}.130$
17. $\bar{2}.885$
18. $\bar{2}.982$
19. $\bar{3}.497$
20. $\bar{4}.360$

Exercise 101
1. $\bar{2}.03$
2. $\bar{0}.20$
3. $\bar{4}.81$
4. $\bar{2}.05$
5. $\bar{6}.89$
6. 3.60
7. $\bar{2}.29$
8. 3.08
9. $\bar{5}.93$
10. $\bar{3}.78$
11. $\bar{4}.092$
12. $\bar{5}.734$
13. $\bar{5}.815$
14. $\bar{5}.930$
15. $\bar{2}.935$
16. $\bar{1}.640$
17. $\bar{1}.308$
18. $\bar{1}.465$
19. $\bar{1}.888$
20. $\bar{2}.921$

Exercise 102
1. 0.0841
2. 0.000253
3. 0.0000174
4. 0.401
5. 0.00120
6. 0.000476
7. 0.585
8. 0.0000172
9. 0.0233
10. 0.535
11. 0.121
12. 0.00176
13. 0.0286
14. 0.00411
15. 0.0000207
16. 0.252
17. 0.00000238
18. 0.000379
19. 0.00183
20. 0.124

Exercise 103
1. 0.150
2. 3.95
3. 5.64
4. 0.0798
5. 0.000789
6. 2.89
7. 0.514
8. 0.00701
9. 0.430
10. 17.3

Exercise 104
1. 0.137
2. 0.0798
3. 1.30
4. 146.
5. 2920.
6. 124.
7. 6.12
8. 5.96
9. 0.00675
10. 0.00407

Exercise 105
1. 0.0203
2. 0.714
3. 0.00233
4. 0.151
5. 0.189
6. 0.0000163
7. 0.0000518
8. 0.000313
9. 0.233
10. 0.340

Exercise 106
1. 0.281
2. 0.540
3. 0.411
4. 0.262
5. 0.899
6. 0.314
7. 0.0494
8. 0.630
9. 0.813
10. 0.448

Exercise 107
1. 0.156
2. 0.126
3. 0.141
4. 282.
5. 2260.
6. 0.167
7. 0.00362
8. 0.918
9. 0.00162
10. 0.0594

Exercise 108
1. 648 m^2
2. 129.6 m^2; £15.55
3. 25.2 and 25200 cm^2
4. 432 ares
5. 539.55 ares
6. 8.4 m^2
7. 103.2 m^2; £17.54
8. 22.737 m^2; £95.50
9. 143 ares; £44.33
10. 62.25 m long
    21.75 m broad
    2407 slabs
    £770.24 cost

Exercise 109
1. 5.2 m
2. 32 m
3. 120 m
4. 8.7 m
5. Length 120 cm
   Breadth 40 cm
   48 tiles used

Exercise 110
1. 306 cm^2
2. 3.045 cm^2
3. 42.5 cm^2
4. 57.12 cm^2
5. 19.98 cm^2
6. 72.24 cm^2
7. 1.84 m^2
8. 4.085 m^2
9. 262.625 cm^2
10. 7.56 m^2

Exercise 111
1. 62.8 m^2
2. 12.4 m
3. 140.1 m^2
4. 50.22 cm^2
5. 3.6 m^2
6. 732 ares
7. $116\frac{7}{8}$ cm^2
8. 5.4 m^2
9. 4.03 m^2
10. 1932.9 m^2
11. 80.435 cm^2
12. 31.185 cm^2
13. 37.50 cm^2
14. 48.125 cm^2
15. 1.3 m^2
16. 106.24 cm^2
17. 105.16 cm^2
18. 645 m^2
19. 102.18 m^2; £17.37
20. 422.5 cm^2

Exercise 112
1. 22 cm
2. 33 cm
3. 176 cm
4. 61.6 cm
5. 29.7 m
6. 16.5 cm
7. 110 m
8. 5.5 cm
9. 198 m
10. 48.4 cm

Exercise 113
1. 18.84 cm
2. 31.4 cm
3. 62.8 cm
4. 28.26 cm
5. 40.82 cm
6. 50.24 m
7. 75.36 cm
8. 11.932 m
9. 21.038 m
10. 25.434 cm

Exercise 114
1. £22.44
2. 1500 revolutions
3. 30000 revolutions 52.8 Km/hr
4. 400 m
5. Length 21.12 m Breadth 14.08 m

Exercise 115
1. 22¾ cm
2. 1.575 m
3. 35 cm
4. 2.1 m
5. 110¼ cm
6. 1.4 m
7. 10.5 cm
8. 5.6 m
9. 9.8 cm
10. 3.15 m

Exercise 116
1. 346.5 m²
2. 962.5 cm²
3. 37.5 m²
4. 5544 cm²
5. 394.24 cm²
6. 6.16 m²
7. $17\frac{1}{9}$ m²
8. $26\frac{53}{72}$ m²
9. 221.76 cm²
10. $2165\frac{5}{8}$ cm²

Exercise 117
1. 1.98 m
2. 29.7 cm; 187 cm
3. 34.6 cm
4. 24.7 m; 155.1 m
5. 8.8 cm
6. 2.9 m; 5.8 m
7. 6.2 m; 19.5 m
8. 60.4 cm
9. 2.5 m; 7.8 m
10. 21.6 cm

Exercise 118
1. 61.6; 28.4 m²
2. 714 tops; 3127¾ cm²
3. 9.625 m²
4. 346½ m² ; 66 m
5. 44.275 m²

Exercise 119
1. 4.68 m³ ; 17.08 m²
2. 129600 cm³
3. 1.152 m³
4. 2304 litres; 10.56 m²
5. 52000 litres; 8 tins
6. $4666\frac{2}{3}$ litres
7. 22500 bricks
8. 311.04 m³
9. 226.8 m³
10. 7.8 m² ; 2016 l

Exercise 120
1. 5 m
2. 75000 l
3. 1.80 m
4. 64 cm
5. 10 m; 24 cm

Exercise 121
1. 196 cm³
2. 315 cm³
3. 110.4 cm³
4. 126 cm³
5. 4.5 cm³
6. 51.7 cm³
7. 3.024 m³
8. 2.4 m; 1.536 m³
9. 5 cm
10. 4.2 cm

Exercise 122
1. 15.57 cm²; 171.27 cm³
2. 750 cm³
3. 2925 m³
4. 9.45 m³
5. 10631.25 m³

Exercise 123
1. 6688.2 cm²; 39600 cm³
2. 332.64 cm³; 158.4 cm²
3. 346500 cm³; 27500 cm²
4. 2165.6 cm³
5. 9504 m³; 144 l
6. $233\frac{1}{3}$ mins
7. 6160 cm³
8. 2/3 cm
9. 6 cm
10. 147.84 m²
11. 1039.5 cm³; 403.92 cm²; 51975 l; 2019.6 m²
12. 42778 l
13. 33900 cm³
14. 0.85 m
15. 137.5 cm³

Exercise 124
1. 792 cm³
2. 1540 cm³
3. 27.104 cm³
4. 4583 l
5. 8.04 cm³
6. 21 cm
7. 10.5 cm
8. 9 cm
9. 48 cm
10. Pyramid 1960 cm³; Cone 1540 cm³; Difference 420 cm³

Exercise 125
1. 616 cm²; $1437\frac{1}{3}$ cm³
2. 5024 cm²; $33493\frac{1}{3}$ cm³
3. 1026.7 l
4. $718\frac{2}{3}$ cm³
5. 190896 litres
6. 42 litres of paint
7. 904.32 cm³; 452.16 cm²
8. 6.2 cm
9. 8850 balls
10. 10.6 cm; 1410 cm²

Exercise 126
1. £1472; £88
2. £136.50; £2476.50
3. £105.30
4. £78
5. 240 shares; £50.40
6. £566.40; £57.60
7. £2250 invested
8. Increase of £7.20
9. $73\frac{1}{3}$p per share
10. £24 increase

Exercise 127
1. £510 and £30
2. £862.50; £60
3. £535; £30
4. £600; £30
5. £1898; £109.50
6. £126
7. £3300 stock; £231
8. £4500 stock; £247.50
9. £7200 stock; £576
10. £9300 stock; £418.50

Exercise 128
1. £405 increase
2. £27.60 increase
3. £2625; £800; £52
4. £20 income; £250 in second stock; £4056 in first stock; £3380 in second stock
5. £750 of new stock. Increase in income of £10
6. £8400 from sale Loses £10 by reinvesting
7. £346.80
8. £1.50
9. £5000 stock; £262.50; £275 gain
10. £382.50 income; Increase of £22.75

Exercise 129
1. E
2. A
3. A
4. E
5. E
6. A
7. E
8. E
9. A
10. A

Exercise 130
1. 1 second
2. 1/100 cm²
3. 1 day
4. 1 Km
5. 1 litre
6. 1/10 hectare
7. 1/10 quintal
8. 1/10 second
9. 1/100 tonne
10. 1/1000 Km

Exercise 131
1. 0.5 mm
2. 0.05 seconds
3. 0.5 kg
4. 0.05 l
5. 0.005 g
6. 0.05 ares
7. 0.5 day
8. 0.05 Km
9. 0.0005 cm
10. 0.005 kg

Exercise 132
1. 65.5 and 64.5 seconds
2. 1.645 and 1.635 seconds
3. 68.45 and 68.35 cm
4. 0.3755 and .3745 mm
5. 9.85 and 9.75 l
6. 24.935 and 24.925 cm³
7. 2.75 and 2.65 kg
8. 0.955 and 0.945 g
9. 36.55 and 36.45 ares
10. 94.75 and 94.64 m²

Exercise 133
1. 0.011 and 1.1%
2. 0.01 and 1%
3. 0.056 and 5.6%
4. 0.021 and 2.1%
5. 0.0068 and 0.68%
6. 0.0037 and 0.37%
7. 0.042 and 4.2%
8. 0.0049 and 0.49%
9. 0.0077 and 0.77%
10. 0.0062 and 0.62%

Exercise 134
1. 0.3 litres
2. 0.5 cm
3. 0.2 seconds
4. 0.02 g
5. 0.04 mm

Exercise 135
1. 9.4 and 9.0 seconds
2. 57.1 and 56.5 mm
3. 442 and 438 m
4. 0.69 and 0.67 seconds
5. 8.47 and 8.41 g

Exercise 136
1. 1 m 53.2 s
   1 m 53.4 s
2. 89.75 and 89.45 m
3. 16.15 and 15.85 cm
4. 159.0 litres
5. 62.9; 62.5 seconds
6. 251 m; 249 m; 10.7 sec; 10.9 sec; 23.46 and 22.84 m/sec
7. 236 and 228 cm
8. 81.75 hrs
9. 79.55 cm; 77.65 cm
10. 13500; 13627.75; 13372.75 and 4209.25 cm^2

Exercise 137
1. 1/6; 5/36; 13/18; 1/6
2. 3/8; 1/8
3. 1/52; 1/13; 1/26; 1/2; 1/13
4. 1/20; 6/125
5. 1/221; 11/221; 1/17; 25/102
6. 1/169; 1/4; 1/208
7. 2/5; 1/5; 3/10
8. 30/119; 60/119 190/1309
9. 1/8
10. 11/52; 13/104

Exercise 138
1. 16 arrangements 3/8; 1/16; 1/2
2. 10 arrangements
3. 18 different menus
4. 10 in preliminary round
   1st round, 2nd round, semi final and final
5. 35 different ways 1/35

Exercise 139
1. 12
2. 3^8
3. 8 x 7 x 6 = 336
4. 56
5. 1295
6. 608400
7. 120, 40, 20
8. 840, 120, 400, 240, 40
9. 120
10. 64
11. 336
12. 120
13. 6
14. 90
15. 240
16. 119
17. 120, 325
18. 2520
19. 50
20. 60, 504

Exercise 141
1. D
2. C
3. C
4. D
5. C
6. C
7. C
8. D
9. C
10. D

Exercise 142

1.
Score	Frequency
1	1
2	5
3	7
4	14
5	17
6	23
7	17
8	8
9	6
10	2

Modal Score is 6.

2.
No. of Bottles	Frequency
1	34
2	33
3	24
4	8
5	3
6	2

Modal Number is 1 bottle.

3.
No. of Children	Frequency
0	3
1	14
2	22
3	25
4	17
5	11
6	6
7	2

Modal size of family is 3 children.

4.
Total	Frequency
2	2
3	7
4	11
5	15
6	18
7	17
8	9
9	12
10	5
11	2
12	2

Mode is 6.

5.
Shoe Size	Frequency
2	4
3	8
4	16
5	21
6	29
7	18
8	9
9	7
10	3

Modal size of shoe is 6.

Exercise 143

1. | Score | Frequency |
|---|---|
| 1-10 | 2 |
| 11-20 | 10 |
| 21-30 | 18 |
| 31-40 | 16 |
| 41-50 | 21 |
| 51-60 | 24 |
| 61-70 | 18 |
| 71-80 | 6 |
| 81-90 | 3 |
| 91-100 | 2 |

Modal class is 6th i.e. 51-60.

2. | Length | Frequency |
|---|---|
| 20.5-21.5 | 2 |
| 21.5-22.5 | 6 |
| 22.5-23.5 | 17 |
| 23.5-24.5 | 41 |
| 24.5-25.5 | 24 |
| 25.5-26.5 | 7 |
| 26.5-27.5 | 3 |

Modal class 23.5-24.5 cm

3. | Mark | Frequency |
|---|---|
| 35-39 | 5 |
| 40-44 | 6 |
| 45-49 | 9 |
| 50-54 | 13 |
| 55-59 | 17 |
| 60-64 | 10 |
| 65-69 | 8 |
| 70-74 | 6 |
| 75-79 | 6 |

Modal class is 55-59

4. | Time | Frequency |
|---|---|
| 1-5 min | 1 |
| 6-10 min | 3 |
| 11-15 min | 9 |
| 16-20 min | 12 |
| 21-25 min | 18 |
| 26-30 min | 20 |
| 31-35 min | 13 |
| 36-40 min | 11 |
| 41-45 min | 8 |
| 46-50 min | 5 |

Modal class is 26-30 min. 6th class boundaries 25.5-30.5 min.

Exercise 144

1. 100 pupils
2. Modal class 51-60. Third class limits 31-40. Sixth class boundaries 60.5-70.5. 140 pupils were tested. Probability is 33/140.
3. 59%, 70 kg or more.
4. | No. of Eggs | Frequency |
|---|---|
| 1 | 24 |
| 2 | 14 |
| 3 | 23 |
| 4 | 29 |
| 5 | 9 |
| 6 | 11 |
| 7 | 5 |
| 8 | 5 |

5. From length of words, book would be a childs.

Exercise 145

1. 65
2. 63.8 Km
3. 17.9°C.
4. 5 min. 14 secs.
5. 30 kg

Exercise 146

1. 4 articles. Probability $\frac{17}{75}$.
2. 9 questions. 15.6% fail
3. 40.23 g mean weight. 140 pkts tested. 36% underweight.
4. 6.5 books
5. (i) 6.9
   (iii) a) 1/12
   b) 1/6
   c) 1/18
   (iv) a) 16
   b) 33
   c) 11

Exercise 147
1. 59.4
2. 67.1 kg
3. 10.71 seconds
4. Boys percentage absences 12.8%. Girls percentage absences 11.2%. Girls have best attendance record.
5. 17 litres.

Exercise 148
1. Lower quartile 43.
   Median 56.
   Upper quartile 72.
2. Lower quartile 166.
   Median 175.
   Upper quartile 184.
3. Lower quartile 17.
   Median 19.5.
   Upper quartile 22.
4. Lower quartile 322.
   Median 346.
   Upper quartile 363.
5. Lower quartile 39.
   Median 46.
   Upper quartile 53.

Exercise 149
1.
Shoe Size	Cumulative Frequency
2	7
3	22
4	53
5	89
6	131
7	164
8	180
9	194
10	200

$Q_1 = 4$; $Q_2 = 6$; $Q_3 = 7$.
Mean size 6.

2.
Upper Class Boundary	Cumulative Frequency
20.5	5
30.5	15
40.5	40
50.5	90
60.6	143
70.5	203
80.5	287
90.5	349
100.5	380

$Q_1 = 51.5$; $Q_2 = 68.5$; $Q_3 = 70.5$.
Semi-interquartile range = 9.5.
Mean = 65.7
46.6% will obtain an A pass.

3.
Weight	Cumulative Frequency
56	2
57	7
58	14
59	22
60	42
61	55
62	69
63	79
64	89
65	93
66	96
67	98
68	100

$Q_1 = 59.1$ g; $Q_2 = 60.5$ g; $Q_3 = 62.6$ g.
22% underweight;
2000 weight 60 g.

contd. overleaf

4. Mean = 47.2 mm
    Q_1 = 41.7 mm
    Q_2 = 47.6 mm
    Q_3 = 53 mm
Semi-interquartile range
= 5.6 mm.

Upper Boundaries	Cumulative Frequency
29.5	5
34.5	18
39.5	52
44.5	107
49.5	180
54.5	242
59.5	287
64.5	297
69.5	300

5.

Upper Boundaries	Cumulative Frequency
220.5 l	7
240.5 l	42
260.5 l	141
280.5 l	288
300.5 l	469
320.5 l	829
340.5 l	1080
360.5 l	1247
380.5 l	1313
400.5 l	1344
420.5 l	1350

1350 householders investigated.
Q_1 = 285 l
Q_2 = 312 l
Q_3 = 335 l
Semi-interquartile range 25 litres.
46800000 litres.